T0305715

Computational Framework for the Finite Element Method in MATLAB® and Python

Computational Framework for the Finite Element Method in MATLAB® and Python aims to provide a programming framework for coding linear FEM using matrix-based MATLAB® language and Python scripting language. It describes FEM algorithm implementation in the most generic formulation so that it is possible to apply this algorithm to as many application problems as possible.

Readers can follow the step-by-step process of developing algorithms with clear explanations of its underlying mathematics and how to put it into MATLAB and Python code. The content is focused on aspects of numerical methods and coding FEM rather than FEM mathematical analysis. However, basic mathematical formulations for numerical techniques which are needed to implement FEM are provided. Particular attention is paid to an efficient programming style using sparse matrices.

Features

- Contains ready-to-use coding recipes allowing fast prototyping and solving of mathematical problems using FEM
- Suitable for upper-level undergraduates and graduates in applied mathematics, science or engineering
- Both MATLAB and Python programming codes are provided to give readers more flexibility in the practical framework implementation

Pavel Sumets is a research and development software engineer from New Zealand whose main interest and expertise are biomechanics, numerical methods and scientific programming.

Computational Framework for the Finite Element Method in MATLAB® and Python

Pavel Sumets

CRC Press
Taylor & Francis Group
Boca Raton London New York

CRC Press is an imprint of the
Taylor & Francis Group, an **informa** business

A CHAPMAN & HALL BOOK

First edition published 2023
by CRC Press
6000 Broken Sound Parkway NW, Suite 300, Boca Raton, FL 33487-2742

and by CRC Press
4 Park Square, Milton Park, Abingdon, Oxon, OX14 4RN

ISBN: 978-1-032-20925-8 (hbk)
ISBN: 978-1-032-20927-2 (pbk)
ISBN: 978-1-003-26597-9 (ebk)

DOI: 10.1201/9781003265979

Typeset in LM Roman font
by KnowledgeWorks Global Ltd.

Contents

Introduction

I F you are studying or working in the area of engineering, applied mathematics or physics, it is very likely that you have already heard about the finite element method (FEM). Probably, you have read some literature on FEM and have tried to solve your particular problem using some FEM software package. Based on my own experience of doing so, it is not easy for the newcomer to get into the subject matter. Things become more complicated if you are a researcher and try to solve a nonstandard problem, which requires building customized programming code. Apart from the underlying mathematics, implementing FEM involves scientific programming, which includes coding various numerical methods such as polynomial interpolation, integrating with quadrature rules and a linear solver for matrix equations. Not to mention that building geometric representation for a particular customized problem is not always a trivial problem, all these tasks form quite a complicated FEM algorithm to be solved. Probably, implementing each individual component of the algorithm does not represent a challenge, but combining them all together in one interconnected system which should work efficiently and correctly is what makes FEM implementation challenging.

When an engineer or researcher faces a need to solve a problem using FEM, basically, there are three possible options: use existing standard FEM software packages, rewrite and optimize some third-party FEM library for particular needs or write everything from scratch. Obviously, if someone writes everything from scratch, the subject knowledge should cover all aspects of an underlying FEM algorithm. The advantage of having comprehensive knowledge about how FEM works is that it allows us to understand any existing FEM software packages as well. The content of this book is written with an idea to give necessary knowledge to develop numerical coding skills and become flexible with solving FEM.

Who this book is intended for:

This book is for those who want to gain skills in scientific programming numerical methods and better understand their advantages and drawbacks. Although the pivot point of the book is the FEM algorithm, various complementary tasks are considered. To be specific, the book provides a programming framework for coding linear FEM algorithms using the matrix-based MATLAB® language and the Python scripting language in application to solving ordinary and partial differential equations. Special attention is paid to building numerical algorithms and its efficient implementation for both one- and two-dimensional problems. A reader can follow a step-by-step process of developing algorithms with explanations of their underlying mathematics and how to put them into MATLAB and Python code with corresponding analysis of their efficiency.

The content is focussed on aspects of numerical methods and coding FEM algorithms rather than FEM mathematical analysis; however, basic mathematical formulations for numerical techniques needed to implement FEM algorithms are provided.

Level of presentation:

The mathematical background is assumed to be covered by matrix algebra, vector calculus and the basics of ordinary and partial differential equations. It is going to be a demonstration of how the FEM works in the one- and two-dimensional case, written as applied mathematicians usually write it. The content throughout the book is not restricted to either Laplace or Poisson equations but rather covers general linear elliptic boundary value problems. Unlike the majority of the books on the FEM method which imply that a reader is familiar with structural mechanics, the current book is based on more generic formulations without assuming any mechanics background. Structural finite elements are not considered in examples and they are not discussed at all since, contrary to expectations, the structural elements are much more complicated to describe and use. They are left to an advanced FEM course. The book has a number of appendices which contain more detailed explanations of the numerical methods used in the main body of the book, so a reader can refresh or enrich their corresponding knowledge.

We use MATLAB and Python as programming languages to code the steps of the FEM algorithm. Throughout the book, the ideas of coding are illustrated using MATLAB with complementary translation to Python as well. Providing the code examples in both programming languages is done to add flexibility in framework implementation.

In terms of programming background, it is expected that a reader is familiar with basic matrix and vector manipulation using MATLAB and the NumPy package in Python. Every code example given in the book is explained in detail so that even a programmer with little experience could understand the underlying algorithm logic and implementation. We pay attention to an efficient programming style using sparse matrices.

Distinctive features of the book:

There is a large number of books on FEM including FEM mathematical analysis, coding FEM and solving application problems using FEM (mostly problems on mechanics). However, not many of them cover both theory and numerical implementation with coding examples. Usually, the programming implementation comes as complementary to the theoretical part just to illustrate implementation for basic simple cases. Very often, programming implementation is provided for specific application problems with tailored code, and the majority of these applications comes from the structural mechanics.

The current book is designed with an idea to describe FEM algorithm implementation itself and to do this in the most generic formulation so as to make it possible to apply this algorithm to as many application problems as possible. Here, the theoretical part is given as much as needed to implement the corresponding numerical methods which are part of the FEM algorithm. The book pays attention to the important practical details which arise when an engineer writes programming code. Overall, the book considers the FEM method from the scientific programming point of view.

Topics covered by the book content:

The content of the book is divided into six chapters and covers the following subjects:

- Lagrange interpolating polynomials;

- quadrature rules for numerical integration;

- defining parameters of the FEM and its influence on solution errors;

- calculating matrices of the FEM;

- assembling matrices of the FEM;

- incorporating boundary conditions into FEM matrices;

- FEM data structures;

- defining geometry and building two-dimensional triangular mesh;

- adaptation to curved boundaries (isoparametric finite elements);

- solving model problems using FEM.

We shall start with formulation of the model boundary value problem for the linear one-dimensional ordinary differential equation. Then, by introducing interpolating functions and finite elements, FEM algorithms for solving model problems are derived. The two-dimensional problem for a linear partial differential equation is considered afterwards and the corresponding two-dimensional FEM algorithms are derived. Each step of the algorithms is provided in detail with programming function implementations. We discuss data structures to represent mesh triangulation and boundary conditions, and compare different approaches of the matrices assembling process. Full programming code is given for an example problem with validating computational efficiency.

The source code for listings in the book is downloadable and can be found in the repository `https://github.com/psumets/FEM_lib`

Background and Basic Concepts

THE finite element method (FEM) is a numerical method to solve ordinary and partial differential equations arising in mathematical modelling. Originally developed for solving problems in solid-state mechanics, it proved to be efficient and robust in various practical applications ranging from traditional fields of structural analysis, heat transfer, fluid flow, mass transport and electromagnetic potential, to simulating geometry deformation in computer character animation.

The fundamental idea of the FEM is subdividing the computational domain into smaller and simpler parts that are called finite elements, and finding local solutions within the boundary of these elements. By stitching the individual solutions on final elements back together, a global solution can be obtained. Some of the FEM advantages are the ability of accurate representation of complex geometry, capture of local effects and easy solution representation.

The history of the method is quite instructive. FEM concepts were initially introduced by Courant in 1943 in his work on solving the torsion problem, but this work did not attract much attention. Later, in the 1950s, structural engineers put the FEM on the map as a practical technique for solving their elasticity problem. The method had now become an important technique from both the practical and theoretical point of view, and the number of the method's applications began to increase at a tremendous rate. In the 1970s, mathematicians gave a comprehensive account of the method and its underlying theory. By the current time, the FEM has evolved into a universal tool for solving differential equations numerically. Numerous software packages adapted FEM to use with applications in all areas of computational physics and engineering. ANSYS, ADINA and LS-DYNA are examples of general-purpose computer codes for industrial users of the FEM.

Any numerical modelling of a phenomenon starts with formulating a *mathematical model* to be solved. Then, a *numerical model* is constructed that can be solved using computers. The final stage is development of a *computer model* (computer code) to simulate behaviour of the system. The FEM represents a numerical model, so it describes how to transform model equations, geometry and boundary conditions into a discretized form suitable to be consumed by numerical methods. An original mathematical model could have various forms and the resulting numerical model varies as well, but the FEM approach dictates a certain tactic of obtaining a numerical model and it follows certain basic concepts. As such, algorithms of the FEM can be considered as universal tools to approach numerical modelling. Finally, a computer model represents a software package which usually has a form of library with computer implementation of numerical methods that make up FEM algorithms (for example, numerical integration or a linear solver). Building a library code is an art of scientific programming which implies not only good knowledge of the mathematics behind the numerical methods but good programming skills as well. Choice of programming language affects all of the development process as every language has its own unique features and it could come with existing numerical libraries which can be reused or tailored for use with the FEM. Also, very often the code should be optimized to have good performance with the lowest possible calculation error. In case of FEM, a numerical model is often modified slightly so as to take shape of an optimized algorithm which is ready to be coded.

Let us outline the main steps of solving a problem using FEM. As a starting point, we have a mathematical model formulated in terms of a differential equation along with geometry and boundary conditions. First, this model needs to be converted in the form of integral equations with integrals over a geometry domain (in some cases, the original mathematical model is already expressed in an integral form). Once the integral formulation is obtained, the next step is to perform discretization when continuous variables or functions are transformed into discrete counterparts. Discretization includes splitting the geometry domain into individual small patches (final elements) and representing initially continuous functions and integrals as combinations of its interpolation over finite elements. In other words, the solution is sought in the form of piecewise interpolation and coefficients of this interpolation are unknowns which can be found from the discrete model representation. Usually, after discretization, the model transforms into a system of equations which could be either linear or nonlinear, depending on the initial form of the differential equation. The final step is to solve the system and represent the final solution through a piecewise interpolating polynomial.

Programming FEM involves coding various numerical methods and binding them together into one pipeline. It is worth noting that every piece of the pipeline matters as any error introduced at any stage propagates further with accumulative effect which could invalidate the final result. That is why coding numerical implementation is an important part of numerical simulation which should not be underestimated. At a high level, coding includes the following main stages: Firstly, an algorithm of geometry discretization comes into play where final elements are generated. Then, based on the shape and distribution of these elements, implementation of particular numerical integration methods has to be coded. The next important part is to define an algorithm of assembling all these calculated values over individual finite elements into a matrix which forms forming linear system to be solved numerically.

Once the FEM algorithm is coded, validation is carried out to make sure that the implementation is correct. This is usually done against a benchmark with a known analytical solution so that we can estimate error and convergence of the method with various discretization parameters.

There are a large number of books on FEM including FEM mathematical analysis, coding FEM and solving application problems using FEM which could be helpful for understanding FEM algorithms. For example, a reader could find a comprehensive FEM analysis in [1] which covers finite elements of different types. The book by [3] is focussed on steady-state boundary value problems. It contains thorough theoretical and practical aspects of implementing FEM and explains how to write a finite element code from scratch. MATLAB implementation of FEM algorithms with detailed descriptions of numerical analysis and meshing techniques can be found in [5] and [2]. Reading [4] and [7] is useful to learn about using Python for scientific computing. For better understanding of how to use specific MATLAB tools to solve differential equations, see documentation [6].

CHAPTER 1

Finite Element Method for the One-Dimensional Boundary Value Problem

IT was mentioned in the introduction of basic concepts that the development of the finite element method (FEM) in practical terms was established by engineers to solve their structural problems. For this reason, the FEM is usually approached through considering displacement of a system of linear springs under applied force. This technique is developed from the stiffness approach applied to structural problems and equations arising from this technique are force-displacement relationships which model specific mechanics. Unlike this conventional approach, the method of introducing FEM in this book is based on broader formulation considering a generic differential equation without any structural mechanics reference. All particular problems, including structural ones, can be modelled as a special case of this generic equation.

In this chapter we obtain FEM algorithm for linear one-dimensional second-order ordinary differential equation (ODE) with mixed boundary conditions. First, we derive integral equation for the boundary value problem and outline basic approach of the FEM. Then, finite elements

DOI: 10.1201/9781003265979-1

and basis functions are introduced along with polynomial interpolation techniques. We show how the integral equation can be discretized with finite elements and how the FEM is reduced to solving linear system.

GLOSSARY

$u, v, \ldots$: vectors are written using bold fonts.

$(u \cdot v)$: dot product of two vectors, $u \cdot v = \sum\limits_{i=1}^{n} u_i v_i$.

$U, V, \ldots$: matrices are denoted by capital letters.

$f(x)$: a function of a single variable.

$f', f'', \ldots$: derivatives of a single variable function $f(x)$.

$C(a, b)$: set of continuous on the interval (a, b) functions.

$C^k(a, b)$: set of continuous functions that have continuous first k derivatives.

$L_2(a, b)$: space of functions defined on (a, b) for which the second power of the absolute value is Lebesgue integrable. If $v(x) \in L_2(a, b)$, then $\|v\|_{L_2}^2 = \int_a^b |v|^2 dx < \infty$.

$H^1(a, b)$: Sobolev space defined as $H^1(a, b) := \{v \in L_2(a, b) : v' \in L_2(a, b)\}$. If $v(x) \in H^1(a, b)$, then $\|v\|_{H^1}^2 = \int_a^b \left(|v'|^2\right) dx < \infty$.

$\mathcal{G}_m$: set of polynomials of degree not greater than m.

1.1 FORMULATION OF THE BOUNDARY VALUE PROBLEM FOR LINEAR SECOND-ORDER ODE

Throughout the chapter, we consider linear ODE of the form

$$- \left(c(x)u'(x)\right)' + b(x)u'(x) + a(x)u(x) = f(x), \quad x \in (s_0, s_1). \quad (1.1)$$

Here, $c(x), b(x), a(x)$ and $f(x)$ are known functions; $u(x)$ is unknown function to be found, and $u' = du/dx$. It is assumed that $c \in C^1[s_0, s_1]$; $b, a, f \in C[s_0, s_1]$ and $c(x) > 0 \ \forall x \in [s_0, s_1]$. Negative sign in equation (1.1) is for convenience and the reason of keeping this sign will be clarified later. We are going to derive FEM algorithm, which is based on equation (1.1).

By considering equation of the form (1.1), we cover the problem in most generic formulations which allows solving specific problems as well; for example, when $c(x) = -1$, $b(x) = a(x) = 0$, we come to Poisson equation, $u''(x) = f(x)$. Another example could be one-dimensional steady-state heat equation with heat sink, $-(c(x)u'(x))') + u(x) = 0$, where $u(x)$ is a temperature and $c(x)$ is a thermal conductivity. This equation is a special case of more generic one (1.1).

Note that an ODE allows multiple solutions and addition conditions are needed to obtain unique solution. To find a unique solution to equation (1.1), two boundary conditions are needed (second-order equation needs two extra conditions to be uniquely resolved). If both boundary conditions are specified at a single point (either s_0 or s_1), then this problem is called *Cauchy problem*. Problem (1.1) with boundary conditions specified at both, s_0 and s_1, boundary points is called *boundary value problem* (BVP). Cauchy problem and BVP have distinct properties and solution methods. This book is focussed on BVP only.

Let us consider various types of conditions, which can be specified at a boundary point s (here, s could be any of $\{s_0, s_1\}$). For example, solution is known at point s with $u(s) = u_s$, which gives us Dirichlet (or first-type) boundary condition. Another situation is when a derivative of solution at boundary point is defined $u'(s) = u'_s$ which is called Neumann (or second-type) boundary condition. Combination of Dirichlet and Neumann boundary conditions gives Robin (or third-type) boundary condition having form $u'(s) + \rho u(s) = \mu$, or $\pm(cu')(s) + \sigma u(s) = \mu$ with corresponding choice of σ and μ (here, function $c(x)$ is a coefficient of equation (1.1)). Usually, these conditions have some meaning depending on particular application problem. For example, in application to the heat equation with $u(x)$ being a temperature, Dirichlet boundary condition means that a temperature at a boundary point is known, and Neumann boundary condition specifies heat flux at a boundary point.

If boundary conditions (BC) of different types are given for a BVP, then it is called that a mixed boundary condition is defined. For example, problem (1.1) can be complemented by BC

$$u(s_0) = u_{s_0}, \quad (cu')(s_1) + \sigma u(s_1) = \mu. \tag{1.2}$$

Here, Dirichlet BC is specified at s_0 and Robin BC is given at s_1 which all together give mixed BC. Further, conditions (1.2) are used as a BC to find a unique solution to the problem (1.1). Now, let us give a definition of the solution function $u(x)$.

Definition 1.1 *Function $u(x) \in C^2(s_0, s_1) \cap C^1(s_0, s_1] \cap C[s_0, s_1]$ satisfying equation (1.1) and BC (1.2) is called a classical solution to the BVP.*

It is assumed that there exists a unique solution $u(x)$, and further we are focussed on solving BVP (1.1)–(1.2) using FEM. To reach this goal, we need to transform the problem to an integral equation which is shown in the next section.

EXERCISES

▶ To see clearly how solution property depends on particular BC, solve the equation

$$u'' + 3u' - 4u = 0, \quad x \in (0, 1)$$

with two sets of BC:

$$\text{(a)} \quad u(0) = 1, u'(0) = 0;$$

$$\text{(b)} \quad u(0) = 1, u(1) = 0.$$

Case (a) is Cauchy problem having solution, which increases exponentially while solution corresponding to case (b) decreases.

▶ Unlike Cauchy problem, even a simple BVP could have no solution. Explore the problem:

$$u'' + u = 0, \quad x \in (0, \pi), \quad u(0) = 1, \quad u(\pi) = \alpha,$$

where α is a parameter. Find values of α for which solution exists. Is this solution unique?

▶ Find solution to the problem

$$-u'' + u = 0, \quad x \in (0, \pi), \quad u(0) = 1, \quad u(\pi) = \alpha,$$

and show that it has unique solution for any value of parameter α.

1.2 INTEGRAL EQUATION

Let us transform the problems (1.1)–(1.2) into an integral equation suitable for applying FEM. There are several approaches of doing this, which include variational and projective methods. In this book we use projective method while variational approach is discussed in Appendix A. Projective method is based on the orthogonal L_2 projection on the Sobolev space, $H^1(s_0, s_1)$.

Definition 1.2 *Given a function* $f(x) \in L_2(s_0, s_1)$ *and a space* $H^1(s_0, s_1)$, *the* L_2 *orthogonal projection* $Pr(f) \in H^1(s_0, s_1)$ *of a function* $f(x)$ *is defined by*

$$\int_{s_0}^{s_1} (f(x) - Pr(f)) \, v(x) dx = 0, \quad \forall v(x) \in H^1(s_0, s_1). \quad (1.3)$$

In analogy with projection onto subspaces of $\mathbb{R}^n$, (1.3) defines a projection of f onto H^1, since the difference $f - Pr(f)$ is required to be orthogonal to all functions v in H^1 (see figure 1.1).

From Definition 1.2, it follows that L_2 projection can be viewed as a technique for approximating functions, which gives a weighted average approximation of function f by its projection and weights $v(x)$

$$\int_{s_0}^{s_1} f(x)v(x)dx = \int_{s_0}^{s_1} Pr(f)v(x)dx, \quad \begin{aligned} & Pr(f) \in H^1(s_0, s_1), \\ & \forall v(x) \in H^1(s_0, s_1). \end{aligned} \quad (1.4)$$

In this context, the function $v(x)$ is referred to as a weight or *test function*. Now, considering f in (1.4) to be right hand side of equation (1.1),

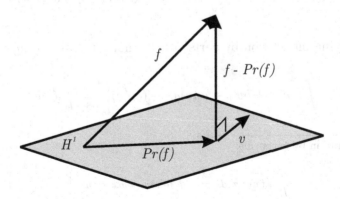

FIGURE 1.1: Illustration of L_2 orthogonal projection.

projection $Pr(f)$ corresponds to projecting left hand side of equation (1.1) on to the space $H^1(s_0, s_1)$, as a result we obtain an integral expression

$$\int_{s_0}^{s_1} \left((-c(x)u'(x))' + b(x)u'(x) + a(x)u(x) \right) v(x)dx$$

$$= \int_{s_0}^{s_1} f(x)v(x)dx, \quad (1.5)$$

where $u(x)$, $v(x) \in H^1(s_0, s_1)$. The intuition behind representing original BVP (1.1) in the integral form (1.5) can be viewed as follows: If $u(x)$ were an exact solution over the whole interval (s_0, s_1), equation (1.1) would be exact everywhere. But, given that in real engineering problems this will not be the case, we try to obtain an approximate solution $u(x)$ for which it meets the BVP equation in a spatially averaged sense (1.5). In other words, the residual or error (i.e., the amount by which the differential equation is not satisfied exactly at a point) is distributed evenly over the interval.

Let us rewrite expression (1.5) in the form with integrals over individual terms

$$-\int_{s_0}^{s_1} (c(x)u'(x))' v(x)dx + \int_{s_0}^{s_1} b(x)u'(x)v(x)dx$$

$$+ \int_{s_0}^{s_1} a(x)u(x)v(x)dx = \int_{s_0}^{s_1} f(x)v(x)dx. \quad (1.6)$$

By applying integration by parts[1] to the first integral in (1.6) we come to

$$\int_{s_0}^{s_1} (cu'v' + bu'v + auv)dx - (cu'v)\Big|_{s_0}^{s_1} = \int_{s_0}^{s_1} fvdx. \quad (1.7)$$

---

[1]Integration by parts rule:

$$\int_{s_0}^{s_1} u'(x)v(x)dx = -\int_{s_0}^{s_1} u(x)v'(x)dx + (uv)\Big|_{s_0}^{s_1}.$$

By inspecting second term in (1.7) and taking into account BC (1.2), we see that

$$(cu'v)\Big|_{s_0}^{s_1} = c(s_1)u(s_1)'v(s_1) - c(s_0)u(s_0)'v(s_0)$$

$$= \mu v(s_1) - \sigma u(s_1)v(s_1) - c(s_0)u(s_0)'v(s_0). \quad (1.8)$$

If function $v(x)$ is defined so that $v(s_0) = 0$, then it follows from (1.7) and (1.8) that

$$\int_{s_0}^{s_1} (cu'v' + bu'v + auv)\, dx + \sigma u(s_1)v(s_1) = \int_{s_0}^{s_1} fv dx + \mu v(s_1). \quad (1.9)$$

Expression (1.9) is an integral equation, and it can be seen that if $u(x)$ is a solution to BVP (1.1)–(1.2), then it satisfies (1.9). Inverse would be valid if we complement integral equation (1.9) by the condition $u(s_0) = u_{s_0}$. As a result, equation (1.1) and BC (1.2) are now transformed into one expression (1.9) where $u(s_0) = u_{s_0}$ and $v(s_0) = 0$. It is worth noting here that original BVP suggests that the solution u should have derivatives up to second order. On the other hand, the integral form (1.9) refers only to the first derivatives of $u(x)$, which implies weaker conditions on property of solution function. Moreover, due to property of Sobolev space, L_2 projection $u(x)$ does not require the function we seek to have continuous derivatives. As such, we relax some conditions on solution of BVP by saying that the function $u(x)$ is now integrable continuous function having piecewise continuous derivatives of the first order (in other words, it belongs to Sobolev space). For this reason, (1.9) is called a *weak form* of the original BVP. The integral weak form of (1.1)–(1.2) can now be defined in terms of the Sobolev space by introducing two subspaces

$$\mathcal{W} := \{w \in H^1(s_0, s_1) : w(s_0) = u_{s_0}\},$$
$$\mathcal{W}^0 := \{w \in H^1(s_0, s_1) : w(s_0) = 0\},$$

as follows.

Definition 1.3 *Expression (1.9) with $u(x) \in \mathcal{W}$ and arbitrary function $v(x) \in \mathcal{W}^0$ is called the integral equation corresponding to the BVP (1.1)–(1.2). Function $u(x)$ is called a weak solution if it satisfies (1.9).*

Obtaining expression (1.9) is the first step in the FEM formulation for the BVP. Next step is to represent solution through polynomial approximation. Since we consider $u(x)$ to be piecewise smooth, piecewise polynomial interpolation is used to approximate $u(x)$. Generally speaking, FEM can be expressed as following: *FEM = integral equation + piecewise polynomial interpolation*. In the next section we explore polynomial interpolation methods.

EXERCISES

▶ Prove that (1.1)–(1.2) follow from (1.9). Instruction: take any point $x_0 \in (s_0, s_1)$; consider an interval $[x_0 - h, x_0 + h] \subset (s_0, s_1)$ and use the expression (1.7) defined on this interval with $v \in C^1[s_0, s_1]$ and $v(x) = 0$ for $x \notin [x_0 - h, x_0 + h]$; perform passage to the limit $h \to 0$ and obtain (1.1).

▶ Deduce integral equation and define sets $\mathcal{W}$ and $\mathcal{W}^0$ for the problem (1.1) with following BC:

  (a) BC of the third order is specified at the point s_0 and BC of the first order at the point s_1.

  (b) BC of the third order is specified at both points, s_0 and s_1.

1.3 LAGRANGE INTERPOLATING POLYNOMIALS AND THE FINITE ELEMENT APPROXIMATION

Let us consider function approximation methods. Here, we discuss two approaches: using single global interpolating polynomial and piecewise polynomial interpolation. Both approaches are based on Lagrange interpolating polynomials. Since the concept of piecewise interpolation is an extension of the global one, we start with considering global interpolation first.

1.3.1 Global Interpolating Polynomial

Let $u = u(x)$ be a continuous on the closed interval $[s_0, s_1]$ function with defined set of nodes

$$s_0 = x_1 < x_2 < \cdots < x_{m+1} = s_1, \quad h = s_1 - s_0,$$

and $u_i = u(x_i), i = 1, \ldots, m+1$. Then, there is a unique polynomial u_h of degree not greater than m satisfying

$$u_h(x_i) = u_i, \quad i = 1, \ldots, m+1. \tag{1.10}$$

Definition 1.4 *The points x_i are called interpolation points or interpolation nodes and expression (1.10) is the interpolation condition. Polynomial $u_h(x)$ is considered to be an approximation of function $u(x)$ on the interval $[s_0, s_1]$.*

Note that $u_h(x)$ is not the same as $u(x)$, but rather an approximation, which coincides with $u(x)$ in a given set of nodes. Polynomial approximation $u_h(x)$ can be created in different ways, but we are interested in Lagrange interpolating polynomials having the form

$$u_h(x) = \sum_{i=1}^{m+1} u_i \varphi_i(x), \quad \varphi_i(x) = \prod_{j=1, j \neq i}^{m+1} \frac{x - x_j}{x_i - x_j}. \tag{1.11}$$

In other words, we seek an approximation in the form of linear combination of the polynomials $\varphi_i(x)$ with known weights u_i.

Definition 1.5 *Polynomials $\varphi_i(x)$ in (1.11) are called Lagrange basis functions.*

It can be seen that

$$\varphi_i \in \mathcal{G}_m, \quad \varphi_i(x_j) = \delta_{ij}, \quad j = 1, \ldots, m+1,$$

where $\mathcal{G}_m$ denotes a set of polynomials of degree not greater than m, and δ_{ij} is the Kronecker delta.[2] Functions $\varphi_i(x)$ are called basis due to the fact that they form a basis in $\mathcal{G}_m$ and any polynomial $p \in \mathcal{G}_m$ can be represented through this basis (see 1.11). Particularly, if $u_h \equiv 1$, then it follows from (1.11) that

$$\sum_{i=1}^{m+1} \varphi_i(x) = 1, \quad x \in [s_0, s_1]. \tag{1.12}$$

---

[2]Kronecker delta function is defined as:

$$\delta_{ij} = \begin{cases} 0, & \text{if } i \neq j, \\ 1, & \text{if } i = j. \end{cases}$$

It is worth noting that evaluating the basis functions $\varphi_i(x)$ using expression (1.11) requires $O(m^2)$ operations which becomes computationally expensive for large values of m. For this reason, it is desired to rewrite (1.11) in another form of interpolation formula, which can be referred as the barycentric formula

$$\varphi_i(x) = w(x)\frac{\beta_i}{x - x_i}, \quad u_h(x) = w(x)\sum_{i=1}^{m+1}\frac{\beta_i}{x - x_i}u_i, \qquad (1.13)$$

where

$$w(x) = \prod_{j=1}^{m+1}(x - x_j), \quad \beta_i = \left(\prod_{j=1,j\neq i}^{m+1}(x_i - x_j)\right)^{-1}, \qquad (1.14)$$

and coefficients β_i are called weights of barycentric formula. It follows from (1.12) that

$$1 = \sum_{i=1}^{m+1}\varphi_i(x) = w(x)\sum_{i=1}^{m+1}\frac{\beta_i}{x - x_i},$$

and by using this fact, we come to the final form of the barycentric interpolation formula

$$u_h(x) = \left(\sum_{i=1}^{m+1}\frac{\beta_i}{x - x_i}\right)^{-1}\sum_{i=1}^{m+1}\frac{\beta_i}{x - x_i}u_i, \qquad (1.15)$$

and basis functions

$$\varphi_i(x) = \left(\sum_{i=1}^{m+1}\frac{\beta_i}{x - x_i}\right)^{-1}\frac{\beta_i}{x - x_i}. \qquad (1.16)$$

It is easy to see that calculations of $u_h(x)$ using expression (1.15) for any given x require $O(m)$ operations. Weights β_i do not depend on x and ones precomputed can be reused for each evaluation of $u_h(x)$.

An important property of representations (1.15)–(1.16) is that they remain valid when weights β_i are multiplied by an arbitrary constant value. The weights in the denominator are exactly the same as in the numerator, and this means that any common factor in all the weights may be cancelled. As such, weights β_i can be defined up to constant factor.

For particular case of nodes distribution, explicit formulas for barycentric weights can be given. For instance, in case of evenly spaced points $x_{i+1} - x_i = h/m$, weights can be calculated through formula

$$\beta_i = (-1)^i \frac{m!}{(m - i + 1)!(i - 1)!}, \quad i = 1, \ldots, m + 1. \qquad (1.17)$$

To illustrate calculation of weights β_i, let us consider 4 nodes $\{x_1 = 0, x_2 = 1, x_3 = 2, x_4 = 3\}$. Using expression (1.14), we come to values $\{\beta_1 = -1/6, \beta_2 = 1/2, \beta_3 = -1/2, \beta_4 = 1/6\}$, while formula (1.17) gives $\{\beta_1 = -1, \beta_2 = 3, \beta_3 = -3, \beta_4 = 1\}$, which differ by factor 6; nevertheless, they can be equally used in (1.15)–(1.16). Figure 1.2 illustrates four basis functions created from the given four nodes and each basis is a polynomial of the third order. Note that using this basis allows to interpolate exactly any cubic polynomial (and polynomial of degree less than three), which is shown in figure 1.3.

Accuracy and some drawbacks of the global approximation are discussed in Appendix B, where it is shown that the strategy of increasing

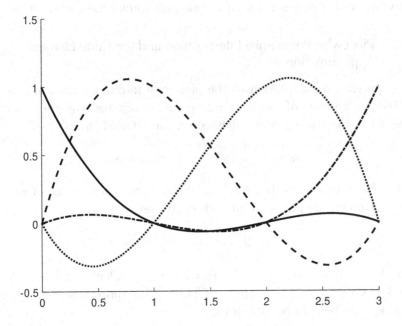

FIGURE 1.2: Lagrange basis functions corresponding to nodes $\{x_1 = 0, x_2 = 1, x_3 = 2, x_4 = 3\}$.

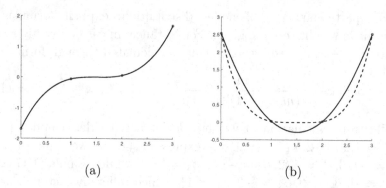

(a)   (b)

FIGURE 1.3: Examples of interpolations with four nodes corresponding to functions: (a) $u(x) = 0.5(x - 1.5)^3$; (b) $u(x) = 0.5(x - 1.5)^4$. Here, the dashed line is the source function and the solid line is the interpolation. Cubic function is interpolated exactly; however, the fourth-order polynomial is interpolated exactly only at nodes but not between nodes.

polynomial order to interpolate function having large number of interpolation nodes is not always a good approach. Below, we consider alternative approach representing piecewise polynomial interpolation.

1.3.2 Piecewise Polynomial Interpolation and the Finite Element Approximation

Now, we turn our attention to the piecewise interpolation, which is effectively a number of localized interpolation stitched together. Let us consider a set of $n + 1$ nodes defined on the interval $[s_0, s_1]$

$$s_0 = x_1 < x_2 < \cdots < x_{n+1} = s_1,$$

and subintervals $e_i = [x_i, x_{i+1}]$, $i = 1, \ldots, n$. Then, a set of $m + 1$ different nodes are specified on each e_i so that

$$x_i = x_i^{(1)} < x_i^{(2)} < \cdots < x_i^{(m+1)} = x_{i+1},$$

and the total number of nodes can be easily calculated to be $N = nm + 1$. Given a function $u(x) \in C[s_0, s_1]$, its approximation $u_h(x) \in C[s_0, s_1]$ can be defined as follows:

$$u_h(x) \in \mathcal{G}_m, \quad x \in e_i, \quad i = 1, \ldots, n,$$
$$u_h(x_j) = u(x_j), \quad j = 1, \ldots, N,$$

where $h = \max\{h_i, i = 1, \ldots, n\}$. In other words, $[s_0, s_1] = \bigcup_{i=1}^{n} e_i$ and function $u(x)$ is interpolated on each subinterval e_i through Lagrange polynomial of the mth order.

Definition 1.6 *Approximation $u_h(h)$ is called Lagrange spline, and subintervals e_i are called finite elements.*

We can see that each finite element is associated with continuous polynomial $\mathcal{G}_m$, while the spline itself is a piecewise smooth continuous function, which is smooth on each subinterval e_i and continuous across the interval $[s_0, s_1]$ (see figure 1.4).

Let us introduce a space of Lagrange splines of degree m

$$H_h^m(s_0, s_1) := \{w \in C[s_0, s_1] : w(x) \in \mathcal{G}_m, \ x \in e_i, \ i = 1, \ldots, n\}.$$

Space $H_h^m(s_0, s_1)$ is considered to be an approximation of the space $H^1(s_0, s_1)$, and $H_h^m(s_0, s_1) \subset H^1(s_0, s_1)$. Indeed, if $w(x) \in H_h^m(s_0, s_1)$

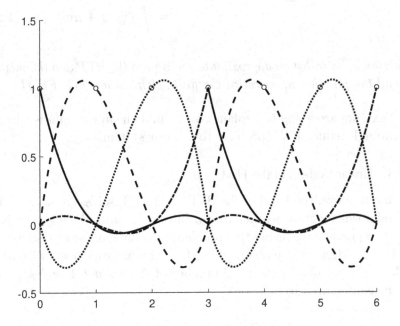

FIGURE 1.4: Lagrange basis functions corresponding to the spline with two finite elements $e_1 = [0, 3]$ and $e_2 = [3, 6]$ having nodes $\{x_1 = 0, x_2 = 1, x_3 = 2, x_4 = 3, x_5 = 4, x_6 = 5, x_7 = 6\}$. It is seen that spline at the node $x_4 = 3$ is continuous but not smooth function.

then $w(x)$ is a continuous and piecewise differentiable function which means that $w \in H^1(s_0, s_1)$. Analogously, approximations of the subspaces $\mathcal{W}$ and $\mathcal{W}^0$ associated with integral equation (1.8) can be introduced.

$$\mathcal{W}_h := \{w \in H_h^m(s_0, s_1) : w(s_0) = u_{s_0}\},$$
$$\mathcal{W}_h^0 := \{w \in H_h^m(s_0, s_1) : w(s_0) = 0\}.$$

Now, we have come to the state when we are ready to give a definition of the FEM corresponding to the BVP (1.1)–(1.2).

Definition 1.7 *The finite element approximation of the BVP (1.1)–(1.2) is defined through the integral equation (1.8) where $u = u_h \in \mathcal{W}_h$ and $v = v_h \in \mathcal{W}_h^0$.*

$$\int_{s_0}^{s_1} \left(cu_h' v_h' + bu_h' v_h + a u_h v_h\right) dx + \sigma u_h(s_1) v_h(s_1)$$

$$= \int_{s_0}^{s_1} f v_h dx + \mu v_h(s_1). \quad (1.18)$$

Function u_h is called an approximate solution to the BVP and the method to find the function u_h is called the finite element method (FEM).

Next, we are going to explain how to find the approximate solution u_h through reducing the problem to a linear system.

1.3.3 Linear System of the FEM

As shown above, the FEM of the BVP (1.1)–(1.2) solves equation (1.18) for unknown function u_h and with arbitrary v_h. Introducing a basis in $H_h^m(s_0, s_1)$ allows reducing this problem to solving a linear system.

By considering the interval $[s_0, s_1]$ having n finite elements with $m+1$ nodes each, we obtain the total number of $N = nm + 1$ nodes, which can be enumerated from 1 to N:

$$s_0 = x_1 < x_2 < \cdots < x_N = s_1.$$

Each node x_i can be associated with a Lagrange basis function $\varphi_i(x) \in H_h^m(s_0, s_1)$. Basis introduced in section 1.3.1 comprises smooth

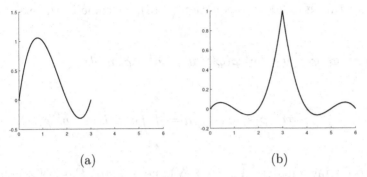

(a)   (b)

FIGURE 1.5: Examples of basis functions associated with Lagrange spline having two finite elements $e_1 = [0, 3]$, $e_2 = [3, 6]$, and seven nodes $\{x_1 = 0, x_2 = 1, \ldots, x_7 = 6\}$: (a) basis $\varphi_2(x)$ corresponding to the node $x_2 = 1$; (b) basis $\varphi_4(x)$ corresponding to the node $x_4 = 3$.

functions while $u_h \in H_h^m(s_0, s_1)$ is a piecewise smooth Lagrange spline and corresponding basis comprises piecewise smooth functions as well. Functions $\varphi_i(x)$ have a property of taking nonzero value only in the corresponding finite element interval and zero otherwise. A basis function associated with a node, which is shared by two finite elements, is nonzero in these two elements (see figure 1.5). As such, every $\varphi_i(x)$ takes nonzero values in either one or two finite elements. Recall that $\varphi_i(x_j) = 0$ if $i \neq j$, and $\varphi_i(x_i) = 1$.

Let us denote function node values as $u_i = u_h(x_i)$ and $v_i = v_h(x_i)$, then any functions $u_h \in \mathcal{W}_h$ and $v_h \in \mathcal{W}_h^0$ can be decomposed in terms of basis $\varphi_i(x) \in H_h^m(s_0, s_1)$ as following [3]

$$u_h(x) = \sum_{j=1}^{N} u_j \varphi_j(x) = u_{s_0} \varphi_1(x) + \sum_{j=2}^{N} u_j \varphi_j(x),$$

$$v_h(x) = \sum_{i=1}^{N} v_i \varphi_i(x), \quad x \in [s_0, s_1],$$

(1.19)

or

$$u_h(x) = \boldsymbol{\varphi}^T \boldsymbol{u}, \quad v_h(x) = \boldsymbol{\varphi}^T \boldsymbol{v},$$

(1.20)

where $\boldsymbol{u} = \{u_i\}_{i=1}^{N}$, $\boldsymbol{v} = \{v_i\}_{i=1}^{N}$ and $\boldsymbol{\varphi} = \{\varphi_i(x)\}_{i=1}^{N}$ are column-vectors.

---

[3]Note that $v_0 = 0$ by definition of subspace $\mathcal{W}_h^0$ to which v_h belongs.

By substituting (1.20) into equation (1.18), we come to the expression[4]

$$\int_{s_0}^{s_1} \left(c \boldsymbol{v}^T \boldsymbol{\varphi}'(\boldsymbol{\varphi}')^T \boldsymbol{u} + b \boldsymbol{v}^T \boldsymbol{\varphi}(\boldsymbol{\varphi}')^T \boldsymbol{u} + a \boldsymbol{v}^T \boldsymbol{\varphi}\boldsymbol{\varphi}^T \boldsymbol{u} \right) dx$$

$$+ \sigma \boldsymbol{v}^T \boldsymbol{\varphi}(s_1)\boldsymbol{\varphi}(s_1)^T \boldsymbol{u} = \int_{s_0}^{s_1} f \boldsymbol{v}^T \boldsymbol{\varphi} dx + \mu \boldsymbol{v}^T \boldsymbol{\varphi}(s_1). \quad (1.21)$$

By introducing a matrix $\tilde{A} \sim (N \times N)$ and a vector $\tilde{\boldsymbol{F}} \sim (N \times 1)$ having components

$$\tilde{A}_{ij} = \int_{s_0}^{s_1} \left(c \varphi_i' \varphi_j' + b \varphi_i \varphi_j' + a \varphi_i \varphi_j \right) dx + \sigma \varphi_i(s_1)\varphi_j(s_1), \quad (1.22)$$

$$\tilde{F}_i = \int_{s_0}^{s_1} f \varphi_i dx + \mu \varphi_i(s_1), \quad (1.23)$$

we can rewrite (1.21) in the form

$$\boldsymbol{v}^T \tilde{A} \boldsymbol{u} = \boldsymbol{v}^T \tilde{\boldsymbol{F}}. \quad (1.24)$$

Expression (1.24) can be seen as a sum of equations defined by the system $\tilde{A}\boldsymbol{u} = \tilde{\boldsymbol{F}}$ with each equation being multiplied by v_i (actually, we can ignore the first equation since it corresponds to the value $v_1 \equiv 0$). Identity (1.24) is valid for arbitrary $\boldsymbol{v}$, so by cancelling it out and by splitting the vector $\boldsymbol{u}$ into unknown and known parts, we come to the linear system

$$\sum_{j=2}^{N} \tilde{A}_{ij} u_j = \tilde{F}_i - u_{s_0} \tilde{A}_{i1}, \quad i = 2, \dots, N. \quad (1.25)$$

From solving linear system (1.25), we obtain values for u_j and the final approximate solution to the BVP can be found using the first formula in (1.19).

---

[4]Here, we use the fact that

$$u_h(x)v_h(x) = \boldsymbol{\varphi}^T \boldsymbol{u}\boldsymbol{\varphi}^T \boldsymbol{v} = \boldsymbol{v}^T \boldsymbol{\varphi}\boldsymbol{\varphi}^T \boldsymbol{u}.$$

Definition 1.8 *Matrix $\tilde{A}$ is called a global stiffness matrix and $\tilde{F}$ is a global forcing vector.*[5] *The system of linear equations (1.25) is called a linear system of the FEM.*

Elements of the stiffness matrix and forcing vector are calculated using expressions (1.22) and (1.23); however, it is impractical to use these formulas directly. Below, we consider efficient method of assembling matrices and derive corresponding computational algorithm for calculating their elements.

EXERCISES

▶ Plot the Lagrange basis functions corresponding to the spline with final elements $e_1 = [0, 1]$, $e_2 = [1, 2]$ and $e_3 = [2, 3]$ having nodes $\{x_1 = 0, x_2 = 1, x_3 = 2, x_4 = 3\}$ (linear basis functions).

▶ Deduce a linear system of the FEM corresponding to the following BC:

(a) BC of the first order is specified at both points, s_0 and s_1.

(b) BC of the third order is specified at both points, s_0 and s_1.

1.4 ILLUSTRATIVE PROBLEM

One of the reasons of the success of the FEM is that it may be applied to problems of great complexity; however, to use such problems for illustrative purpose tends to obscure the underlying ideas. In this section a simple problem is solved. Note that this is not to suggest that the FEM is the best method for solving such problems. In fact, there are other numerical methods available which are superior. However, this particular problem involves only small amount of algebra, and the algorithm ideas come through quite well.

Consider the BVP having the form

$$-u'' + u = x, \quad x \in (0, 1), \quad u(0) = 1, \quad u'(1) = 1, \tag{1.26}$$

with known exact solution

$$u = \frac{e^{2-x} + e^x}{1 + e^2} + x. \tag{1.27}$$

---

[5]These naming come from the stiffness method (also known as the displacement, method), which is used in matrix analysis of structures.

It can be seen from comparing with generic BVP (1.1)–(1.2) that the problem (1.26) corresponds to the coefficient values $c(x) = 1$, $b(x) = 0$, $a(x) = 1$, $\sigma(x) = 0$, $\mu(x) = 1$. Consider a three-element discretization of $[0, 1]$ with nodes $\{x_1 = 0, x_2 = 1/3, x_3 = 2/3, x_4 = 1\}$.

$$e_1 = \left[0, \frac{1}{3}\right], \quad e_2 = \left[\frac{1}{3}, \frac{2}{3}\right], \quad e_2 = \left[\frac{2}{3}, 1\right]. \qquad (1.28)$$

There are four basis functions associated with four interpolation nodes which gives the following form of the approximate solution

$$u_h = u(0)\varphi_1(x) + u\left(\frac{1}{3}\right)\varphi_2(x) + u\left(\frac{2}{3}\right)\varphi_3(x) + u(1)\varphi_4(x). \qquad (1.29)$$

Function value $u(0) = 1$ is known from the BC and the rest values, $u(1/2)$, $u(2/3)$ and $u(1)$ have to be found.

In each element there are just two nodes; hence, the interpolation polynomials for each element must be linear. Basis functions associated with each node are found from (1.11) and have the form

$$\varphi_1(x) = \begin{cases} -3x + 1, & 0 \leq x \leq \frac{1}{3}, \\ 0, & \frac{1}{3} < x < 0, \end{cases} \qquad (1.30)$$

$$\varphi_2(x) = \begin{cases} 3x, & 0 \leq x \leq \frac{1}{3}, \\ -3x + 2, & \frac{1}{3} \leq x \leq \frac{2}{3}, \\ 0, & \frac{2}{3} < x < 0, \end{cases} \qquad (1.31)$$

$$\varphi_3(x) = \begin{cases} 3x - 1, & \frac{1}{3} \leq x \leq \frac{2}{3}, \\ -3x + 3, & \frac{2}{3} \leq x \leq 1, \\ 0, & 1 < x < \frac{1}{3}, \end{cases} \qquad (1.32)$$

$$\varphi_4(x) = \begin{cases} 3x - 2, & \frac{2}{3} \leq x \leq 1, \\ 0, & 1 < x < \frac{2}{3}. \end{cases} \qquad (1.33)$$

Basis functions are shown in figure 1.6 and they have so-called hat shapes. Note that φ_1 and φ_4 are nonzero over one element while φ_2 and φ_3 span two finite elements.

Now, elements of stiffness matrix (1.22) and forcing vector (1.23) have to be found to solve linear FEM system (1.25). First, consider the stiffness matrix in which elements are defined through

$$\tilde{A}_{ij} = \int_0^1 \left(\varphi_i'\varphi_j' + \varphi_i\varphi_j\right) dx, \quad i, j = 1, 2, 3, 4. \qquad (1.34)$$

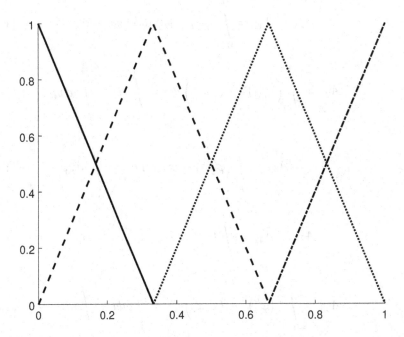

FIGURE 1.6: Lagrange linear basis functions corresponding to the spline with three finite elements $e_1 = [0, 1/3]$, $e_2 = [1/3, 2/3]$ and $e_3 = [2/3, 1]$ having nodes $\{x_1 = 0, x_2 = 1/3, x_3 = 2/3, x_4 = 1\}$.

It is worth looking into expressions for every element to better understand patterns and matrix formation.

$$\tilde{A}_{11} = \int_0^1 \left((\varphi_1')^2 + (\varphi_1)^2 \right) dx = \int_0^{\frac{1}{3}} \cdot dx, \qquad (1.35)$$

$$\tilde{A}_{12} = \tilde{A}_{21} = \int_0^1 \left(\varphi_1' \varphi_2' + \varphi_1 \varphi_2 \right) dx = \int_0^{\frac{1}{3}} \cdot dx, \qquad (1.36)$$

$$\tilde{A}_{13} = \tilde{A}_{31} = \int_0^1 \left(\varphi_1' \varphi_3' + \varphi_1 \varphi_3 \right) dx = 0, \qquad (1.37)$$

$$\tilde{A}_{14} = \tilde{A}_{41} = \int_0^1 \left(\varphi_1'\varphi_4' + \varphi_1\varphi_4\right) dx = 0, \qquad (1.38)$$

$$\tilde{A}_{22} = \int_0^1 \left((\varphi_2')^2 + (\varphi_2)^2\right) dx = \int_0^{\frac{1}{3}} \cdot dx + \int_{\frac{1}{3}}^{\frac{2}{3}} \cdot dx, \qquad (1.39)$$

$$\tilde{A}_{23} = \tilde{A}_{32} = \int_0^1 \left(\varphi_2'\varphi_3' + \varphi_2\varphi_3\right) dx = \int_{\frac{1}{3}}^{\frac{2}{3}} \cdot dx, \qquad (1.40)$$

$$\tilde{A}_{24} = \tilde{A}_{42} = \int_0^1 \left(\varphi_2'\varphi_4' + \varphi_2\varphi_4\right) dx = 0, \qquad (1.41)$$

$$\tilde{A}_{33} = \int_0^1 \left((\varphi_3')^2 + (\varphi_3)^2\right) dx = \int_{\frac{1}{3}}^{\frac{2}{3}} \cdot dx + \int_{\frac{2}{3}}^1 \cdot dx, \qquad (1.42)$$

$$\tilde{A}_{34} = \tilde{A}_{43} = \int_0^1 \left(\varphi_3'\varphi_4' + \varphi_3\varphi_4\right) dx = \int_{\frac{2}{3}}^1 \cdot dx, \qquad (1.43)$$

$$\tilde{A}_{44} = \int_0^1 \left((\varphi_4')^2 + (\varphi_4)^2\right) dx = \int_{\frac{2}{3}}^1 \cdot dx. \qquad (1.44)$$

We can see that due to the symmetry of formula (1.34), index order is not important and the stiffness matrix becomes symmetrical. In addition, elements $\tilde{A}_{13}$, $\tilde{A}_{31}$, $\tilde{A}_{14}$, $\tilde{A}_{41}$, $\tilde{A}_{24}$ and $\tilde{A}_{42}$ equal to zero since corresponding, basis functions are non-overlapping, which gives zeroth integrands (the only coupling occurs between nodes associated with the same element). Also, the domain integrals over the interval $[0,1]$ can be replaced by the sum of integrals taken separately over the finite elements and only those integrals can be left which have non-zero values. Elements $\tilde{A}_{22}$ and $\tilde{A}_{33}$ include two integrals because corresponding integrands are non-zero over two finite elements. As a result, the stiffness matrix can schematically be represented as follows:

$$
\tilde{A} = \begin{pmatrix} * & * & 0 & 0 \\ * & * & * & 0 \\ 0 & * & * & * \\ 0 & 0 & * & * \end{pmatrix} = \int_{0}^{\frac{1}{3}} \cdot dx + \int_{\frac{1}{3}}^{\frac{2}{3}} \cdot dx + \int_{\frac{2}{3}}^{1} \cdot dx
$$

$$
= \begin{pmatrix} * & * & 0 & 0 \\ * & * & 0 & 0 \\ 0 & 0 & 0 & 0 \\ 0 & 0 & 0 & 0 \end{pmatrix} + \begin{pmatrix} 0 & 0 & 0 & 0 \\ 0 & * & * & 0 \\ 0 & * & * & 0 \\ 0 & 0 & 0 & 0 \end{pmatrix} + \begin{pmatrix} 0 & 0 & 0 & 0 \\ 0 & 0 & 0 & 0 \\ 0 & 0 & * & * \\ 0 & 0 & * & * \end{pmatrix} \quad (1.45)
$$

$$
= \bar{A}^{e_1} + \bar{A}^{e_2} + \bar{A}^{e_3}.
$$

Here, $\bar{A}^{e_1}$, $\bar{A}^{e_2}$ and $\bar{A}^{e_3}$ are local to each finite element stiffness matrices, which can be found separately and then summed up to obtain the matrix $\tilde{A}$. One more important observation from (1.45) is that matrices are sparse and $\tilde{A}$ is a tridiagonal matrix.

Now, consider matrix $\bar{A}^{e_1}$ and calculate its first element integral

$$
\bar{A}^{e_1}_{11} = \int_{0}^{\frac{1}{3}} \left(((-3x+1)')^2 + (-3x+1)^2 \right) dx = \left| \begin{matrix} 3x = \hat{x}, \ dx = \frac{1}{3}d\hat{x}, \\ \frac{d}{dx} = \frac{d}{d\hat{x}}\frac{d\hat{x}}{dx} = 3\frac{d}{d\hat{x}} \end{matrix} \right.
$$

$$
= \int_{0}^{1} \left((3(1-\hat{x})')^2 + (1-\hat{x})^2 \right) \frac{1}{3}d\hat{x} = \int_{0}^{1} \left(9 + (1-\hat{x})^2 \right) \frac{1}{3}d\hat{x} = \frac{28}{9}.
$$

$$(1.46)$$

Here, we introduce a new variable, $\hat{x} = 3x$, so that $x \in [0, 1/3] \to \hat{x} \in [0, 1]$, which results in basis functions to be $\varphi_1(\hat{x}) = 1-\hat{x}$ and $\varphi_2(\hat{x}) = \hat{x}$. This transformation, as we see later, allows simplifying calculation of integrals and applying unified numerical integration methods (quadrature rules). Similarly, integrals $\bar{A}^{e_1}_{12}$ and $\bar{A}^{e_1}_{22}$ can be found over $\hat{x}$ space.

$$
\bar{A}^{e_1}_{12} = \int_{0}^{\frac{1}{3}} \left((-3x+1)'(3x)' + (-3x+1)(3x) \right) dx
$$

$$
= \int_{0}^{1} \left(3(1-\hat{x})'3\hat{x}' + (1-\hat{x})\hat{x} \right) \frac{1}{3}d\hat{x} = \int_{0}^{1} \left(-9 + (1-\hat{x})\hat{x} \right) \frac{1}{3}d\hat{x} = -\frac{53}{18}.
$$

$$(1.47)$$

$$
\bar{A}^{e_1}_{22} = \int_{0}^{1} \left((3\hat{x}')^2 + \hat{x}^2 \right) \frac{1}{3}d\hat{x} = \int_{0}^{1} \left(9 + \hat{x}^2 \right) \frac{1}{3}d\hat{x} = \frac{28}{9}. \quad (1.48)
$$

Likewise, change of variables for the element e_2, $\hat{x} = 3x - 1$, results in $x \in [1/3, 2/3] \rightarrow \hat{x} \in [0, 1]$ and the basis functions are $\varphi_2(\hat{x}) = 1 - \hat{x}$, $\varphi_3(\hat{x}) = \hat{x}$. One important observation is that in the space $\hat{x}$ basis functions $\varphi_2(\hat{x})$ and $\varphi_3(\hat{x})$ over the element e_2 are identical to the functions $\varphi_1(\hat{x})$ and $\varphi_2(\hat{x})$ over the element e_1. This leads to the fact that the integrals over the element e_2 are expected to be similar to those over the element e_1

$$\bar{A}_{22}^{e_2} = \int_0^1 \left((3(1 - \hat{x})')^2 + (1 - \hat{x})^2 \right) \frac{1}{3} d\hat{x} = \int_0^1 \left(9 + (1 - \hat{x})^2 \right) \frac{1}{3} d\hat{x} = \frac{28}{9}.$$

$$(1.49)$$

$$\bar{A}_{23}^{e_2} = \int_0^1 \left(-9 + (1 - \hat{x})\hat{x} \right) \frac{1}{3} d\hat{x} = -\frac{53}{18}. \qquad (1.50)$$

$$\bar{A}_{33}^{e_2} = \int_0^1 \left(9 + \hat{x}^2 \right) \frac{1}{3} d\hat{x} = \frac{28}{9}. \qquad (1.51)$$

Finally, for the last element e_3 we have $\hat{x} = 3x - 2$ and $x \in [2/3, 1] \rightarrow \hat{x} \in [0, 1]$ and $\varphi_3(\hat{x}) = 1 - \hat{x}$, $\varphi_4(\hat{x}) = \hat{x}$. Again, $\varphi_3(\hat{x})$ and $\varphi_4(\hat{x})$ over e_3 are identical to $\varphi_2(\hat{x})$ and $\varphi_3(\hat{x})$ over element e_2.

$$\bar{A}_{33}^{e_3} = \int_0^1 \left(9 + (1 - \hat{x})^2 \right) \frac{1}{3} d\hat{x} = \frac{28}{9}. \qquad (1.52)$$

$$\bar{A}_{34}^{e_3} = \int_0^1 \left(-9 + (1 - \hat{x})\hat{x} \right) \frac{1}{3} d\hat{x} = -\frac{53}{18}. \qquad (1.53)$$

$$\bar{A}_{44}^{e_3} = \int_0^1 \left(9 + \hat{x}^2 \right) \frac{1}{3} d\hat{x} = \frac{28}{9}. \qquad (1.54)$$

Once the local stiffness matrices $\bar{A}^{e_1}$, $\bar{A}^{e_2}$ and $\bar{A}^{e_3}$ are found, the final stiffness matrix can be assembled

$$
\tilde{A} =
\begin{pmatrix}
\frac{28}{9} & -\frac{53}{18} & 0 & 0 \\
-\frac{53}{18} & \frac{28}{9} & 0 & 0 \\
0 & 0 & 0 & 0 \\
0 & 0 & 0 & 0
\end{pmatrix}
+
\begin{pmatrix}
0 & 0 & 0 & 0 \\
0 & \frac{28}{9} & -\frac{53}{18} & 0 \\
0 & -\frac{53}{18} & \frac{28}{9} & 0 \\
0 & 0 & 0 & 0
\end{pmatrix}
$$
$$
+
\begin{pmatrix}
0 & 0 & 0 & 0 \\
0 & 0 & 0 & 0 \\
0 & 0 & \frac{28}{9} & -\frac{53}{18} \\
0 & 0 & -\frac{53}{18} & \frac{28}{9}
\end{pmatrix}
=
\begin{pmatrix}
\frac{28}{9} & -\frac{53}{18} & 0 & 0 \\
-\frac{53}{18} & \frac{56}{9} & -\frac{53}{18} & 0 \\
0 & -\frac{53}{18} & \frac{56}{9} & -\frac{53}{18} \\
0 & 0 & -\frac{53}{18} & \frac{28}{9}
\end{pmatrix}.
\tag{1.55}
$$

For the forcing vector expression (1.23) gives

$$
\tilde{F}_i = \int_0^1 x\varphi_i dx + \varphi_i(1), \quad i = 1, 2, 3, 4.
\tag{1.56}
$$

Following the same logic as for stiffness matrix, let us consider individual integrals and deduce vector formation pattern.

$$
\tilde{F}_1 = \int_0^1 x\varphi_1 dx + \varphi_1(1) = \int_0^{\frac{1}{3}} x\varphi_1 dx,
\tag{1.57}
$$

$$
\tilde{F}_2 = \int_0^1 x\varphi_2 dx + \varphi_2(1) = \int_0^{\frac{1}{3}} x\varphi_2 dx + \int_{\frac{1}{3}}^{\frac{2}{3}} x\varphi_2 dx,
\tag{1.58}
$$

$$
\tilde{F}_3 = \int_0^1 x\varphi_3 dx + \varphi_3(1) = \int_{\frac{1}{3}}^{\frac{2}{3}} x\varphi_3 dx + \int_{\frac{2}{3}}^1 x\varphi_3 dx,
\tag{1.59}
$$

$$
\tilde{F}_4 = \int_0^1 x\varphi_4 dx + \varphi_4(1) = \int_{\frac{2}{3}}^1 x\varphi_4 dx + 1.
\tag{1.60}
$$

Here, we take into account the fact that only $\varphi_4(1)$ is non-zero with the unity value. By splitting domain integral into individual integrals over the finite elements, we come to the following schematic representation for the forcing term

$$
\tilde{F} = \begin{pmatrix} * \\ * \\ * \\ * \end{pmatrix} = \int_0^{\frac{1}{3}} \cdot dx + \int_{\frac{1}{3}}^{\frac{2}{3}} \cdot dx + \int_{\frac{2}{3}}^{1} \cdot dx + \varphi_4(1)
$$

$$
= \begin{pmatrix} * \\ * \\ 0 \\ 0 \end{pmatrix} + \begin{pmatrix} 0 \\ * \\ * \\ 0 \end{pmatrix} + \begin{pmatrix} 0 \\ 0 \\ * \\ * \end{pmatrix} + \begin{pmatrix} 0 \\ 0 \\ 0 \\ 1 \end{pmatrix} = \bar{F}^{e_1} + \bar{F}^{e_2} + \bar{F}^{e_3} + \begin{pmatrix} 0 \\ 0 \\ 0 \\ 1 \end{pmatrix}.
$$

(1.61)

Vectors $\bar{F}^{e_1}$, $\bar{F}^{e_2}$ and $\bar{F}^{e_3}$ are considered to be local to each element forcing vectors which can be found separately. Note that local vectors are sparse while the vector $\tilde{F}$ is dense. By introducing new variables $\hat{x}$ for each finite element (see above), corresponding vector elements can be calculated as follows:

$$
\bar{F}_1^{e_1} = \int_0^{\frac{1}{3}} x(-3x + 1)dx = \int_0^1 \frac{1}{3}\hat{x}(1 - \hat{x})d\hat{x} = \frac{1}{54}.
$$

(1.62)

$$
\bar{F}_2^{e_1} = \int_0^{\frac{1}{3}} 3x^2 dx = \int_0^1 \frac{1}{3}\hat{x}^2 d\hat{x} = \frac{1}{27}.
$$

(1.63)

$$
\bar{F}_2^{e_2} = \int_{\frac{1}{3}}^{\frac{2}{3}} x(-3x + 2)dx = \int_0^1 \frac{1}{9}(1 + \hat{x})(1 - \hat{x})d\hat{x} = \frac{2}{27}.
$$

(1.64)

$$
\bar{F}_3^{e_2} = \int_{\frac{1}{3}}^{\frac{2}{3}} x(3x - 1)dx = \int_0^1 \frac{1}{9}(1 + \hat{x})\hat{x}d\hat{x} = \frac{5}{54}.
$$

(1.65)

$$
\bar{F}_3^{e_3} = \int_{\frac{2}{3}}^{1} x(-3x + 3)dx = \int_0^1 \frac{1}{9}(2 + \hat{x})(1 - \hat{x})d\hat{x} = \frac{7}{54}.
$$

(1.66)

$$
\bar{F}_4^{e_3} = \int_{\frac{2}{3}}^{1} x(3x - 2)dx = \int_0^1 \frac{1}{9}(2 + \hat{x})\hat{x}d\hat{x} = \frac{4}{27}.
$$

(1.67)

Assembling the final forcing vector gives

$$\tilde{F} = \begin{pmatrix} \frac{1}{54} \\ \frac{1}{27} \\ 0 \\ 0 \end{pmatrix} + \begin{pmatrix} 0 \\ \frac{2}{27} \\ \frac{5}{54} \\ 0 \end{pmatrix} + \begin{pmatrix} 0 \\ 0 \\ \frac{7}{54} \\ \frac{4}{27} \end{pmatrix} + \begin{pmatrix} 0 \\ 0 \\ 0 \\ 1 \end{pmatrix} = \begin{pmatrix} \frac{1}{54} \\ \frac{1}{9} \\ \frac{2}{9} \\ \frac{31}{27} \end{pmatrix}. \tag{1.68}$$

From (1.55) and (1.68), the equations can be written in the form

$$\begin{pmatrix} \frac{28}{9} & -\frac{53}{18} & 0 & 0 \\ -\frac{53}{18} & \frac{56}{9} & -\frac{53}{18} & 0 \\ 0 & -\frac{53}{18} & \frac{56}{9} & -\frac{53}{18} \\ 0 & 0 & -\frac{53}{18} & \frac{28}{9} \end{pmatrix} \begin{pmatrix} 1 \\ u\left(\frac{1}{3}\right) \\ u\left(\frac{2}{3}\right) \\ u(1) \end{pmatrix} = \begin{pmatrix} \frac{1}{54} \\ \frac{1}{9} \\ \frac{2}{9} \\ \frac{31}{27} \end{pmatrix}, \tag{1.69}$$

or

$$\begin{pmatrix} \frac{28}{9} \\ -\frac{53}{18} \\ 0 \\ 0 \end{pmatrix} + u\left(\frac{1}{3}\right) \begin{pmatrix} -\frac{53}{18} \\ \frac{56}{9} \\ -\frac{53}{18} \\ 0 \end{pmatrix} + u\left(\frac{2}{3}\right) \begin{pmatrix} 0 \\ -\frac{53}{18} \\ \frac{56}{9} \\ -\frac{53}{18} \end{pmatrix}$$

$$+ u(1) \begin{pmatrix} 0 \\ 0 \\ -\frac{53}{18} \\ \frac{28}{9} \end{pmatrix} = \begin{pmatrix} \frac{1}{54} \\ \frac{1}{9} \\ \frac{2}{9} \\ \frac{31}{27} \end{pmatrix}. \tag{1.70}$$

Note that BC specifies $u(0) = 1$ and there are only three unknowns left in the system (1.70). Then, moving first vector in (1.70) to the right hand side of the equation and removing first row in the entire system (see (1.25)) we come to the final linear system for the function values

$$\begin{pmatrix} \frac{56}{9} & -\frac{53}{18} & 0 \\ -\frac{53}{18} & \frac{56}{9} & -\frac{53}{18} \\ 0 & -\frac{53}{18} & \frac{28}{9} \end{pmatrix} \begin{pmatrix} u\left(\frac{1}{3}\right) \\ u\left(\frac{2}{3}\right) \\ u(1) \end{pmatrix} = \begin{pmatrix} \frac{55}{18} \\ \frac{2}{9} \\ \frac{31}{27} \end{pmatrix}. \tag{1.71}$$

Solving equations (1.71) gives

$$u\left(\frac{1}{3}\right) = 1.129, \quad u\left(\frac{2}{3}\right) = 1.345, \quad u(1) = 1.646. \tag{1.72}$$

From the expression (1.27) the exact solutions at these points are 1.131, 1.351 and 1.648, respectively. It can be seen that FEM solution approximates the exact one quite well.

1.5 ALGORITHMS OF THE FINITE ELEMENT METHOD

Below, we are going to compose algorithms needed for programming implementation of the FEM using from top to bottom approach with increasing granularity of details. We start with an algorithm for composing system (1.25), then, assembling matrices is considered, and finally, explanation of how to calculate elements of these matrices is given. Each subsequent step in the hierarchy of algorithms increases granularity of the previous one by introducing more details.

1.5.1 Composing Linear System of the FEM

We have seen that solving BVP using FEM effectively comes down to solving the linear system (1.25) which, in turn, is defined by the matrix $\tilde{A}$ and the vector $\tilde{\boldsymbol{F}}$. It is straightforward to create a generic algorithm for composing FEM linear system (1.25).

Algorithm 1 Composing FEM linear system

---

1. Split the interval $[s_0, s_1]$ into n finite elements each having $m + 1$ points.

2. Assemble matrix $\tilde{A} \sim (N \times N)$ and vector $\tilde{\boldsymbol{F}} \sim (N \times 1)$ where $N = nm + 1$.

3. Incorporate Dirichlet BC, $u(s_0) = u_{s_0}$, by doing following:

   - Delete the first row of the matrix $\tilde{A}$ and the first element of the vector $\tilde{\boldsymbol{F}}$.

   - Calculate $\tilde{\boldsymbol{F}} = \tilde{\boldsymbol{F}} - u_{s_0}\tilde{A}_1$, where $\tilde{A}_1$ is the first column of the matrix $\tilde{A}$.

   - Delete the first column of the matrix $\tilde{A}$.

4. Solve the system $\tilde{A}\boldsymbol{u} = \tilde{\boldsymbol{F}}$ and find $N - 1$ node values of u.

5. Append u_{s_0} to the vector $\boldsymbol{u}$.

There are two points in this algorithm which have to be clarified, namely, how to assemble matrices and how to solve a linear system. Regarding a method of solving linear system, we will discuss it later that this is a sparse linear system which possess a unique, well-defined solution, which allows applying any standard linear solver. As to the method of assembling matrices, this is crucial feature in the context of the FEM, since this is a computationally expensive part which affects effectiveness of the entire FEM (especially when N is large). Next we focus on the assembling algorithm.

1.5.2 Assembling Global Matrices

Let us consider the stiffness matrix (1.22). It is seen from (1.22) that the matrix $\tilde{A}$ is a square one and can be viewed as a sum of two matrices, namely $\tilde{A} = A + S$, where

$$A_{ij} = \int_{s_0}^{s_1} \left(c\varphi_i'\varphi_j' + b\varphi_i\varphi_j' + a\varphi_i\varphi_j \right) dx, \qquad i, j = 1, \ldots, N. \qquad (1.73)$$

$$S_{i,j} = \sigma\varphi_i(s_1)\varphi_j(s_1),$$

Here, $N = nm + 1$ and the matrix S is defined by the boundary conditions. Recall that the interval $[s_0, s_1]$ is split by $n+1$ points into n finite elements each having $m + 1$ points so that $[s_0, s_1] = \bigcup_{l=1}^{n} e_l$. As such, the integral in (1.73) can be represented through the sum of local integrals over the finite elements e_l

$$A_{ij} = \sum_{l=1}^{n} \int_{e_l} \left(c\varphi_i'\varphi_j' + b\varphi_i\varphi_j' + a\varphi_i\varphi_j \right) dx. \qquad (1.74)$$

Now, let us consider more closely an integral over a local finite element e_l and determine only those basis functions which are nonzero over the element e_l. It is easy to see that l^{th} element spans the points $x_{m(l-1)+1}, \ldots, x_{ml+1}$, or, by introducing a local numeration index $\beta = 1, \ldots, m+1$ and a global starting index notation $n_l = m(l-1)$, we can write that $x_{n_l+\beta} \in e_l$. As a result, $\varphi_{n_l+\beta}$ are the basis functions associated with e_l which are nonzero over this element. These nonzero values form a matrix $\bar{A}^l \sim ((m+1) \times (m+1))$ with elements

$$\bar{A}_{\alpha\beta}^l = \int_{x_{n_l+1}}^{x_{n_l+m+1}} \left(c\varphi_{n_l+\alpha}'\varphi_{n_l+\beta}' + b\varphi_{n_l+\alpha}\varphi_{n_l+\beta}' + a\varphi_{n_l+\alpha}\varphi_{n_l+\beta} \right) dx,$$

$$\alpha, \beta = 1, \ldots, m+1. \qquad (1.75)$$

Definition 1.9 *Matrix (1.75) is called a local stiffness matrix of the lth finite element.*

Introducing a local stiffness matrix allows us in representing global stiffness matrix through a sum of local ones. By inspecting (1.74) we observe that A_{ij} has zero value if φ_i and φ_j are nonoverlapping functions ($|i - j| > m$), and nonzero value otherwise. Moreover, the sum (1.74) has single nonzero summand if $j \neq j$ ($|i - j| < m$) and two summands if $i = j$. As such, elements A_{ij} can be expressed as

$$A_{ij} = \sum_{l,\alpha,\beta:i=n_l+\alpha,j=n_l+\beta} \bar{A}_{\alpha\beta}^l.$$

To incorporate all zero elements as well, let us extend a local stiffness matrix by introducing matrix $A^l \sim (N \times N)$ which is a matrix with zero value elements except of a block corresponding to the finite element e_l and defined by the local stiffness matrix (1.75)

$$A_{n_l+\alpha,n_l+\beta}^l = \bar{A}_{\alpha\beta}^l, \quad \alpha, \beta = 1, \ldots, m+1. \qquad (1.76)$$

Definition 1.10 *Matrix (1.76) is called an extended local stiffness matrix corresponding to the lth element.*

Summarizing, we can see that the global stiffness matrix, A, can be obtained from summation of the extended local stiffness matrices, A^l, which, in turn, are calculated through the local matrices (1.75).

$$A = \sum_{l=1}^{n} A^l.$$

Analogously, global forcing vector is a sum of two vectors, $\tilde{F} = F + M$, where (see (1.23))

$$F_i = \int_{s_0}^{s_1} f\varphi_i dx, \quad M_i = \mu\varphi_i(s_1), \quad i = 1, \ldots, N. \qquad (1.77)$$

By introducing a local forcing vector, $\bar{F} \sim ((m + 1) \times 1)$, through

$$\bar{F}_\alpha^l = \int_{x_{n_l+1}}^{x_{n_l+m+1}} f\varphi_{n_l+\alpha} dx, \quad \alpha = 1, \ldots, m+1, \qquad (1.78)$$

and extended local forcing vector, $\boldsymbol{F} \sim (N \times 1)$, as following

$$F^l_{n_l+\alpha} = \bar{F}^l_\alpha, \quad \alpha = 1, \ldots, m+1, \tag{1.79}$$

we come to the expression

$$\boldsymbol{F} = \sum_{l=1}^{n} \boldsymbol{F}^l.$$

Note that the extended local forcing vector, $\boldsymbol{F}^l$, contains all zero elements except of values defined by the local forcing vector (1.79).

Matrix S and vector $\boldsymbol{M}$ result from the boundary conditions and they are specified at the single point s_1. Since $\varphi_i(x_j) = \delta_{ij}$ and $s_1 = x_N$, then each of S and $\boldsymbol{M}$ has the single nonzero value, namely, $S_{NN} = \sigma$ and $M_N = \mu$.

As a result, we come to the algorithm for assembling stiffness matrix and forcing vector.

Algorithm 2 Assembling global matrices

---

1. Set $A = 0$ and $\boldsymbol{F} = 0$, where $A \sim (N \times N)$ and $\boldsymbol{F} \sim (N \times 1)$.

2. For each finite element index $l = 1, \ldots, n$ do:

   - Calculate global index $n_l = m(l-1)$.
   - Calculate elements of the local stiffness matrix $\bar{A}^l \sim (m+1, m+1)$.
   - Calculate elements of the local forcing vector $\bar{\boldsymbol{F}}^l \sim (m+1, 1)$.
   - For each local index $\alpha, \beta = 1, \ldots, m+1$ do:

     (a) $A_{n_l+\alpha, n_l+\beta} = A_{n_l+\alpha, n_l+\beta} + \bar{A}^l_{\alpha\beta}$.

     (b) $F_{n_l+\alpha} = F_{n_l+\alpha} + \bar{F}^l_\alpha$.

3. Incorporate BC at the point s_1 by assigning:

   - $A_{NN} = A_{NN} + \sigma$.
   - $F_N = F_N + \mu$.

Later, we create a programming code implementing this algorithm and discuss it in details. One problem which has not been illuminated so far is how to calculate elements of local matrices. Below, we turn our attention to this problem and deduce an algorithm for calculating these elements.

1.5.3 Calculating Elements of the Local Matrices

We start with a simpler case of calculating elements of the local forcing vector defined by the integral (1.78) for the lth finite element. By inspecting expression (1.78) we observe that the integrand depends on the local basis functions $\varphi_{n_l+\alpha}$ having form of Lagrange polynomial

$$\varphi_{n_l+\alpha}(x) = \prod_{j=1, j\neq\alpha}^{m+1} \frac{x - x_{n_l+j}}{x_{n_l+\alpha} - x_{n_l+j}}, \quad x \in [x_{n_l+1}, x_{n_l+m+1}], \quad (1.80)$$

which depends on the global index n_l. We are aimed to transform expression (1.80) to the one which is independent on the global index so that to obtain invariant basis expression across all finite elements. To achieve this, let us introduce a local variable, $\hat{x}$, and transformation as follows:

$$\hat{x} = \frac{x - x_{n_l+1}}{h_l}, \quad x = x_{n_l+1} + \hat{x}h_l,$$

$$dx = h_l d\hat{x}, \quad h_l = x_{n_l+m+1} - x_{n_l+1}. \quad (1.81)$$

It can be seen the expression (1.81) specifies linear transformation so that $x_{n_l+j} \rightarrow \hat{x}_j$, or

$$x_{n_l+1}, \ldots, x_{n_l+m+1} \rightarrow 0 = \hat{x}_1, \ldots, \hat{x}_{m+1} = 1,$$

and $e_l = [x_{n_l+1}, x_{n_l+m+1}] \rightarrow \hat{e} = [0, 1], \; \forall n_l$. As a result, any finite element is transformed into a canonical, or basis, element of the unity length. Moreover, substitution (1.81) into (1.80) gives

$$\varphi_{n_l+\alpha}(x) = \varphi_{n_l+\alpha}(x_{n_l+1} + \hat{x}h_l) = \prod_{j=1, j\neq\alpha}^{m+1} \frac{\hat{x} - \hat{x}_j}{\hat{x}_\alpha - \hat{x}_j} = \hat{\varphi}_\alpha(\hat{x}), \quad (1.82)$$

where basis functions $\hat{\varphi}_\alpha$ are now expressed in terms of the local variable $\hat{x}$. If we denote $\hat{f} = f(x_{n_l+1} + \hat{x}h_l)$, then integral (1.78) can be rewritten as

$$\bar{F}_\alpha^l = h_l \int_0^1 \hat{f}\hat{\varphi}_\alpha d\hat{x}, \quad \alpha = 1, \ldots, m+1. \quad (1.83)$$

Integral (1.83) can be evaluated exactly in a limited number of cases (linear $\hat{\varphi}_\alpha$ or $\hat{f} = const$), and, generally speaking, we have to apply a numerical approximation technique to calculate (1.83). We utilize a quadrature rule[6] which gives

$$\bar{F}_\alpha^l \approx h_l \sum_{k=1}^{q} \hat{\gamma}_k \hat{f}(\hat{d}_k) \hat{\varphi}_\alpha(\hat{d}_k), \quad 0 \le \hat{d}_k \le 1, \quad \alpha = 1, \ldots, m+1, \quad (1.84)$$

where q is a number of quadrature nodes, $\hat{\gamma}_k$ and $\hat{d}_k$ are quadrature weights and nodes, respectively. Currently, we do not look at the construction of a particular rule leaving this task for the next section. The main result is that we have defined a numerical method allowing calculation of the local forcing vector through the formula (1.84).

In a similar way, computation of the local stiffness matrix can be reduced to the numerical calculations through the approximate quadrature rule. By using transformation (1.81) and introducing functions $\hat{a}$, $\hat{b}$ and $\hat{c}$, integral (1.75) reduces to

$$\bar{A}_{\alpha\beta}^l = \int_0^1 (h_l^{-1} \hat{c} \hat{\varphi}_\alpha' \hat{\varphi}_\beta' + \hat{b} \hat{\varphi}_\alpha \hat{\varphi}_\beta' + h_l \hat{a} \hat{\varphi}_\alpha \hat{\varphi}_\beta) d\hat{x}, \quad \alpha, \beta = 1, \ldots, m+1.$$

Here, we use the fact that $\varphi(x)' = \hat{\varphi}(\hat{x})'/h_l$. Thereby,

$$\bar{A}_{\alpha\beta}^l \approx \sum_{k=1}^{q} \hat{\gamma}_k (h_l^{-1} \hat{c} \hat{\varphi}_\alpha' \hat{\varphi}_\beta' + \hat{b} \hat{\varphi}_\alpha \hat{\varphi}_\beta' + h_l \hat{a} \hat{\varphi}_\alpha \hat{\varphi}_\beta)(\hat{d}_k),$$

$$\alpha, \beta = 1, \ldots, m+1. \quad (1.85)$$

Expressions (1.84) and (1.85) define approximate numerical calculations of the local matrices. Intuitively, it is clear that the higher precision of quadrature approximation, the higher accuracy of the FEM. Below, we imply that integral calculations are performed approximately and the symbol '$\approx$' is substituted by '$=$' to avoid complicated notations.

Let us rewrite expressions (1.84) and (1.85) in matrix forms. To

---

[6]A quadrature rule is an approximation of the definite integral of a function, $f(x)$, stated as a weighted sum of function values at specified points, x_i,

$$\int_0^1 f \, dx \approx \sum_{k=1}^{q} \gamma_k f(x_k).$$

achieve this, we introduce complementary matrices $\hat{E}_{k\alpha} = \hat{\varphi}_\alpha(\hat{d}_k)$ and $\hat{D}_{k\alpha} = \hat{\varphi}'_\alpha(\hat{d}_k)$ having dimension $q \times (m+1)$. With the help of these matrices, formula (1.84) for the local forcing vector takes the matrix form

$$\bar{\boldsymbol{F}}^l = \hat{E}^T \hat{\boldsymbol{F}}^l, \quad \hat{\boldsymbol{F}}^l = h_l \left(\hat{\gamma}_1 \hat{f}(\hat{d}_1), \ldots, \hat{\gamma}_q \hat{f}(\hat{d}_q)\right)^T. \tag{1.86}$$

One interesting observation following from (1.86) is that in case of $h_l = const$ $\forall l$, and when distribution of quadrature nodes, $\hat{d}_k$, remains the same across all finite elements, then the local forcing vector is calculated only once (all local vectors are equal to each other).

Now, it is easy to represent calculations of the local stiffness matrix (1.85) in the form of matrix expression

$$\bar{A}^l = \hat{D}^T \hat{C}^l \hat{D} + \hat{E}^T \hat{B}^l \hat{D} + \hat{E}^T \hat{A}^l \hat{E},$$

where

$$\hat{A}^l = h_l \operatorname{diag}\left(\hat{\gamma}_1 \hat{a}(\hat{d}_1), \ldots, \hat{\gamma}_q \hat{a}(\hat{d}_q)\right),$$

$$\hat{B}^l = \operatorname{diag}\left(\hat{\gamma}_1 \hat{b}(\hat{d}_1), \ldots, \hat{\gamma}_q \hat{b}(\hat{d}_q)\right),$$

$$\hat{C}^l = h_l^{-1} \operatorname{diag}\left(\hat{\gamma}_1 \hat{c}(\hat{d}_1), \ldots, \hat{\gamma}_q \hat{c}(\hat{d}_q)\right).$$

Generally speaking, matrices $\hat{E}$ and $\hat{D}$ can be associated with interpolating matrix and differentiating matrix, respectively. Indeed, for the function $u_h(x)$ we have local representation for $x \in e_l$ (see(1.20))

$$u_h(x) = \sum_{\alpha=1}^{m+1} \varphi_{n_l+\alpha}(x) u_{n_l+\alpha} = \sum_{\alpha=1}^{m+1} \hat{\varphi}_\alpha(\hat{x}) u_{n_l+\alpha} = \hat{\boldsymbol{\varphi}}(\hat{x})^T \boldsymbol{u}^l,$$

$$\boldsymbol{u}^l = \{u_{n_l+\alpha}\}_{\alpha=1}^{m+1}, \quad x \in e_l.$$

For any given point $x = d_k \in e_l$ (for example, d_k is a quadrature node), we obtain that $u_h(d_k) = \hat{\boldsymbol{\varphi}}(\hat{d}_k)^T \boldsymbol{u}^l$, where $\hat{\boldsymbol{\varphi}}(\hat{d}_k)^T$ is the kth row of the matrix $\hat{E}$. Consequently, $\boldsymbol{u}_h(\boldsymbol{d}) = \hat{E}\boldsymbol{u}_l$, where $\boldsymbol{d}$ and $\boldsymbol{u}_h$ are the vectors of points and function values to be interpolated. As such, matrix $\hat{E}$ can be viewed as interpolating matrix. By performing similar reasoning we observe, that $\boldsymbol{u}'_h(\boldsymbol{d}) = h_l^{-1} \hat{D}\boldsymbol{u}_l$, and the matrix $\hat{D}$ can be associated with differentiating operation. These matrices can be used not only for calculation of stiffness matrix, but when plotting functions u_h and u'_h.

Algorithm 3 Calculating local matrices

---

1. Calculate the element's length, $h_l = |e_l|$, and a vector of the interpolation nodes on the basis element, $\hat{x}_\alpha = h_l^{-1}(x_{n_l+\alpha}-x_{n_l+1})$, $\alpha = 1, \ldots, m+1$.

2. Specify basis functions:

$$\hat{\varphi}_\alpha(\hat{x}) = \prod_{j=1, j\neq\alpha}^{m+1} \frac{\hat{x} - \hat{x}_j}{\hat{x}_\alpha - \hat{x}_j},$$

3. Specify a vector of weights, $\hat{\gamma} \sim (q \times 1)$, and a vector of nodes, $\hat{d} \sim (q \times 1)$, for the quadrature rule defined on the basis element $(0 \leq \hat{d}_k \leq 1)$.

4. Calculate the vector $d = x_{n_l+1} + h_l\hat{d}$.

5. Calculate matrices and a vector:

$$\hat{E}_{k,\alpha} = \hat{\varphi}_\alpha(\hat{d}_k),$$
$$\hat{D}_{k,\alpha} = \hat{\varphi}'_\alpha(\hat{d}_k),$$
$$\hat{A}^l = h_l \operatorname{diag}\left(\hat{\gamma}_1 a(d_1), \ldots, \hat{\gamma}_q a(d_q)\right)$$
$$\hat{B}^l = \operatorname{diag}\left(\hat{\gamma}_1 b(d_1), \ldots, \hat{\gamma}_q b(d_q)\right),$$
$$\hat{C}^l = h_l^{-1} \operatorname{diag}\left(\hat{\gamma}_1 c(d_1), \ldots, \hat{\gamma}_q c(d_q)\right),$$
$$\hat{F}^l = h_l \left(\hat{\gamma}_1 f(d_1), \ldots, \hat{\gamma}_q f(d_q)\right)^T.$$

6. Calculate the local stiffness matrix and the local forcing vector:

$$\bar{A}^l = \left(\hat{D}^T \hat{C}^l + \hat{E}^T \hat{B}^l\right) \hat{D} + \hat{E}^T \hat{A}^l \hat{E},$$
$$\bar{F}^l = \hat{E}^T \hat{F}^l.$$

Summarizing, we come to the algorithm for calculating elements of the local stiffness matrix and forcing vector corresponding to the lth finite element.

As we can see, this is the very stage where basis functions come into play, and this algorithm depends on the choice of particular quadrature rule. It is worth noting that if the final elements are of the same length having the same quadrature and interpolation node distribution, then matrices $\hat{E}$ and $\hat{D}$ have to be calculated only once.

Quadrature nodes and weights distribution have to meet certain criteria to be suitable for numerical integration with appropriate accuracy. Below, we discuss some strategies of choosing nodes distribution.

1.6   QUADRATURE RULES

As we have seen, calculating elements of FEM matrices includes numerical integration which can be done through applying some quadrature rule.

$$\int_0^1 \hat{f} d\hat{x} \approx \sum_{k=1}^q \hat{\gamma}_k \hat{f}(\hat{d}_k).$$

Choosing particular values of number q, node values, $\hat{d}_k$, and weights, $\hat{\gamma}_k$, results in different quadrature rules. Below, we utilize two types of quadrature, namely, Gaussian quadrature and Lobatto quadrature rule. Appendix C gives some theoretical background for these rules where basic explanations and examples are given. In addition, theory on calculations of weights and nodes using orthogonal polynomials is given in Appendix D. Below, we use expressions without detailed explanations which can be found in the given appendices.

First, we consider Gaussian quadrature for integrals over the interval $[-1, 1]$ which is the most common and well-studied domain of integration. Integration over the interval $[0, 1]$ can be obtained from the integration over the interval $[-1, 1]$ through the transformation $\hat{x} = 1/2 + \tilde{x}/2$, $\tilde{x} \in [-1, 1]$

$$\int_0^1 \hat{f} d\hat{x} = \frac{1}{2} \int_{-1}^1 \tilde{f} d\tilde{x} \approx \frac{1}{2} \sum_{k=1}^q \tilde{\gamma}_k \tilde{f}(\tilde{d}_k) = \sum_{i=1}^q \hat{\gamma}_k \hat{f}(\hat{d}_k),$$

$$\hat{\gamma}_k = \frac{1}{2}\tilde{\gamma}_k, \quad \hat{d}_k = \frac{1}{2} + \frac{\tilde{d}_k}{2}.$$

It is worth mentioning that Gaussian quadrature is a type of quadrature formula, which has the highest order of precision with the given number of nodes, q, which is equal to $2q-1$. Nodes, $\tilde{d}_k$, are the roots of Legendre polynomials, $L_q(\tilde{x})$, of order q. In general, nodes and weights can be calculated using eigenvalue decomposition of the following square matrix

$$Q = \begin{pmatrix} 0 & \beta_1 & & & \\ \beta_1 & 0 & \beta_2 & & \\ & \ddots & \ddots & \ddots & \\ & & \beta_{q-2} & 0 & \beta_{q-1} \\ & & & \beta_{q-1} & 0 \end{pmatrix}, \qquad \begin{matrix} \beta_j = \dfrac{j}{\sqrt{4j^2-1}}, \\[2mm] j = 1, \ldots, q-1. \end{matrix} \qquad (1.87)$$

If $Q = U\Lambda U^{-1}$, where $\Lambda = \mathrm{diag}(\lambda_1, \ldots, \lambda_q)$ is a diagonal matrix of ordered eigenvalues and U is a matrix of corresponding eigenvectors, then Gaussian quadrature nodes are $\tilde{d}_k = \lambda_k$ and the weights are $\tilde{\gamma}_k = 2U_{1k}^2$.

One important observation about Gaussian quadrature is that there is no guarantee that the set of nodes contains boundary points, -1 or 1. In order to include the end points of the integration, Lobatto quadrature can be used

$$\int_{-1}^{1} \tilde{f} d\tilde{x} \approx \tilde{\gamma}_1 \tilde{f}(-1) + \sum_{k=2}^{q-1} \tilde{\gamma}_k \tilde{f}(\tilde{d}_k) + \tilde{\gamma}_q \tilde{f}(-1).$$

This type of quadrature has the precision of order $2q-3$ where quadrature nodes are roots of the polynomial $(1 - \tilde{x}^2)L'_{q-1}(\tilde{x})$. Value of nodes and weights can be found through the eigenvalue decomposition $Q = U\Lambda U^{-1}$ of the matrix

$$Q = \begin{pmatrix} 0 & \beta_1 & & & \\ \beta_1 & 0 & \beta_2 & & \\ & \ddots & \ddots & \ddots & \\ & & \beta_{q-2} & 0 & \beta_{q-3} \\ & & & \beta_{q-3} & 0 \end{pmatrix}, \qquad \begin{matrix} \beta_j = \sqrt{\dfrac{j(j+2)}{(2j+1)(2j+3)}}, \\[2mm] j = 1, \ldots, q-3. \end{matrix}$$

Nodes of Lobatto quadrature are $\tilde{d} = \{-1, \lambda_i, 1\}$ and weights to be

$$\tilde{\gamma} = \left\{ \frac{2}{q^2-q}, \frac{4U_{1i}^2}{3(1-\lambda_i)}, \frac{2}{q^2-q} \right\}, \quad i = 1, \ldots, q-2,$$

where λ_i and U_i are the eigenvalues of matrix Q. Lobatto quadrature nodes can be chosen to be interpolation nodes on the basis element. In this case, $q = m+1$ and $\hat{x}_\alpha = \hat{d}_\alpha = (1 + \tilde{d}_\alpha)/2$.

1.7 DEFINING PARAMETERS OF THE FEM

From the previous sections we have learned that there are three parameters defining particular FEM scheme: the number of finite elements, n, the degree of approximating polynomial, m, and the number of quadrature nodes on the basis element, q. To build an efficient FEM implementation, it is worth analysing influence of these parameters since they affect property of stiffness matrices and accuracy of the FEM. Below we discuss various points which need to be taken into account when creating particular FEM scheme.

Structure of the stiffness matrix. Numbers n and m determine uniquely the structure of the global stiffness matrix, A, in terms of distribution of the non-zero entries. If $n \gg m$, then the matrix A is the sparse one containing large number of zeroth elements. Contrarily, when n is small but m is large, then the matrix A is considered to be dense (see figure 1.7).

The total number of non-zeroth elements is equal to $n(m + 1)^2 - (n - 1)$, and this number estimates computer memory consumption and computational costs to solve linear system. It is seen that costs increase linearly with n but have quadratic growth with m.

Accuracy of the solution and choice of parameters n and m. Numbers n and m influence accuracy of the approximate solution u_h. Let us briefly illustrate this influence. Assuming that numerical integration is not used (all integrals are calculated exactly), it is stated in the FEM

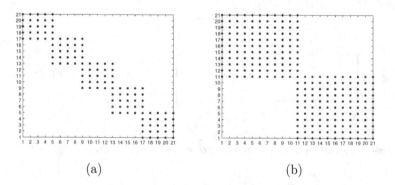

(a)   (b)

FIGURE 1.7: Examples of non-zero entries distribution for the global stiffness matrix: (a) $n = 5$, $m = 4$. Matrix has 121 non-zero elements; (b) $n = 2$, $m = 10$. Matrix has 241 non-zero elements.

theory that

$$\|u - u_h\|_{H^1} \le C \frac{h^{k-1}}{m^{k-1}} \|u\|_{H^k}, \quad \|u\|_{H^k}^2 = \int_{s_0}^{s_1} \left(u^2 + \left| \frac{d^k u}{dx^k} \right|^2 \right) dx.$$

Here C is a constant, u is the exact solution, $2 \le k \le \min\{m + 1, r\}$ where number r is defined from the criteria $\|u\|_{H^r} < \infty$. In connection to this, two cases are of interest to us, namely, (a) solution is not sufficiently smooth function, and $r < (m+1)$; (b) solution is quite smooth function, and $r \gg (m+1)$. By choosing highest value for k, we obtain estimations

$$\text{(a)} \ \|u - u_h\|_{H^1} = O\left(\frac{h^{r-1}}{m^{r-1}} \right), \quad \text{(b)} \ \|u - u_h\|_{H^1} = O\left(\frac{h^m}{m^m} \right).$$

We can see that parameter m affects the solution accuracy differently. To clarify this, let us consider a simplified example when all elements are of the same length, $h = (s_1 - s_0)/n$. In this case, $h = O(n^{-1})$ and we come to the estimations

$$\text{(a)} \ \|u - u_h\|_{H^1} = O\left(\frac{1}{(nm)^{r-1}} \right), \quad \text{(b)} \ \|u - u_h\|_{H^1} = O\left(\frac{1}{(nm)^m} \right).$$

$$(1.88)$$

Consequently, case (a) corresponds to the power growth of accuracy with respect to both parameters, n and m, unlike the case (b) where only increase of n leads to power growth of accuracy while increasing m results in exponential growth.

Thereby, it could be recommended three strategies of how to choose parameters m and n:

1. Having chosen small m (practically, $1 \le m \le 5$), the desired accuracy is reached by increasing n (decreasing h).

2. Having chosen small n (it is allowed to take $n = 1$ which corresponds to the single finite element), the desired accuracy is reached by increasing m (practically, $m = 5, \ldots, 50, \ldots$).

3. Polynomial degree, m, is not fixed for all elements and can vary. This strategy is useful when solution behaves differently on the interval $[s_0, s_1]$. If there is a region with rapidly increasing derivative of the solution, then within this region the value of m is chosen to be fixed and small, but elements are refined (n increases). The value of n in the rest of interval is fixed but the polynomial degree increases.

It is worth noting that the Euclidean condition number of the stiffness matrix, A, is of order $O(h^{-2})$, so refining finite elements is sensible to some extent. Extremely small elements leads to solving ill-conditioned linear system.

In addition, following circumstance has to be taken into account when choosing elements distribution. In practise, differential equation coefficients could be functions having singularities (discontinuities of a function or its derivative) at particular points. In this case, it is advisable to take these points to be boundary nodes of finite elements, that is x_{n_l}. As a result, all functions would be smooth within each finite element and this suffices to obtain desired accuracy of the FEM; otherwise, accuracy decreases significantly.

Accuracy of the solution and distribution of points within finite elements. Vector of points within a basis element $(0 = \hat{x}_1, \ldots, \hat{x}_{m+1} = 1)$ defines distribution of nodes for each finite element. Theoretically (in the exact computer arithmetic), solution u_h does not depend on the way of choosing distribution of these points. Practically (in floating point computation), this distribution plays an important role, particularly, it affects conditional number of the stiffness matrix. When the value of m is small $(m < 7)$, nodes $\hat{x}_i$ can be evenly spaced with a constant step. However, when the value of m is relatively large, it is recommended to choose nodes $\hat{x}_i$ so that they are roots of some orthogonal polynomial of the order $m + 1$. For instance, a good choice would be Chebyshev polynomials or Legendre polynomials.

Quadrature rule and choice of quadrature nodes distribution. Theory states (and results of FEM practical calculation confirm) that the following conditions should be satisfied for a quadrature rule:

- quadrature weights are positive, $\gamma > 0$,

- number of nodes $q \geq m$,

- a quadrature has to be exact for polynomials of degree $2m - 1$ and higher[7]

---

[7]A quadrature rule is called exact for polynomials of degree m, $P_m(x)$, if it yields exact values of integrals for polynomials of that degree

$$\int_{s_0}^{s_1} P_m(x)dx = \sum_{k=1}^{q} \gamma_k P_m(x_k).$$

Practically, there are two quadrature rules meeting these criteria, namely, Gaussian quadrature having number of nodes $q \geq m$, and Lobatto quadrature with $q \geq (m + 1)$. There is a possibility to obtain quadrature nodes, $\hat{d}_i$, which coincide with interpolation nodes on the basis elements, $\hat{x}_i$, and $q = m + 1$. In this case, Lobatto quadrature rule is used for numerical integration since the nodes of this rule include endpoints of the basis element and, in addition, the nodes are roots of orthogonal Legendre polynomial which, in turn, can be taken as nodes of interpolating polynomials on the basis element.

CHAPTER 2

Programming One-Dimensional Finite Element Method for the Linear Boundary Value Problem

I N this chapter, we are going to compose MATLAB® codes for finding numerical solution to the BVP using FEM described in the previous chapter. In addition, algorithm implementation for calculating solution first derivative and solution error will be considered.

Before we start coding FEM algorithm, let us outline the strategy of how it is going to be done. Programming implementation will have a form of library with set of functions with code for particular numerical method or piece of algorithm. We are aimed to group steps of the FEM algorithms into functions according to their logic. For example, from generic high-level FEM algorithm 1 follows that it is convenient to have at least four stand-alone functions for each step of the algorithm, and probably some complementary helper functions as well. The approach we consider below is bottom to top which is opposed to top to bottom consideration in chapter 1. We will start with programming quadrature rules and splitting interval into finite elements; then we will look into coding assembling

DOI: 10.1201/9781003265979-2

stiffness matrices and BC. As a last step, the high-level main function will be considered which implements logic of algorithm 1.

The main data structure used in the code implementation is a matrix, so it is assumed that a reader is familiar with basic skills of matrix manipulation commands in MATLAB and Python. There is, however, a special case of matrices which we will discuss in detail, namely, sparse matrices. As we have seen from chapter 1, FEM results in matrices with sparse structure which we need to reflect in coding.

2.1 SPARSE MATRICES IN MATLAB®

The basic structure to be utilized in the implementation of FEM algorithm is a sparse matrix. The main feature of sparse matrices is the compact data representation when zeroth elements are ignored and not kept in memory. Operations with sparse matrix in MATLAB are similar to operations with full (dense) matrix. The difference is that sparse matrix should be created with specification **sparse()**. There are three methods to create a sparse matrix in MATLAB.

1. Creating a sparse matrix from the full one. In this case the argument of the function **sparse()** is a dense matrix and returning value is a sparse matrix with the shape and values of the dense one. For example,

```
B =[1 3 0
    0 4 0
    2 0 5 ] ;
A = sparse(B)
A = (1,1)    1
    (3,1)    2
    (1,2)    3
    (2,2)    4
    (3,3)    5
```

The sparse matrix is represented in the form of indices of nonzero elements and corresponding values. As a result we see arrays of indices (i,j) and values.

2. Creating a sparse matrix by elements. The previous matrix can be specified as follows

```
A = sparse(3,3) ;
```

```
A(1,1)=1; A(3,1)=2; A(1,2)=3;
A(2,2)=4; A(3,3)=5;
```

Here, we specify matrix dimensions as arguments of the function **sparse()** followed by assigning values of matrix elements.

3. Creating a sparse matrix from the arrays of indices and values. For example, the next code creates the matrix A from previous examples.

```
% indices
i = [ 1 3 1 2 3 ] ;
j = [ 1 1 2 2 3 ] ;
% values
v = [ 1 2 3 4 5 ] ;
% matrix
A = sparse (i,j,v,3,3) ;
```

In general case, values in vectors of indices could repeat and in this case the corresponding values are summed up. This allows formation of the matrix A by accumulating its elements. For example, the matrix A can be specified as

```
i = [ 1 1 3 1 2 3 ] ;
j = [ 1 1 1 2 2 3 ] ;
v = [ 0.3 0.7 2 3 4 5 ] ;
A = sparse (i,j,v,3,3) ;
```

Here A(1,1)=0.3+0.7. It is worth noting that (i,j) could be specified as two-dimensional matrices. In this case, function **sparse()** treats them as column vectors. In addition, the value v could be a scalar and it would be automatically extended to the vector of the corresponding length with all elements of v.

The third method is preferable for creating large sparse matrices since it is more generic and it reflects the way in which matrices are represented in MATLAB internally.

2.2 INPUT DATA STRUCTURES

Let us define input data structure to be used in MATLAB implementation. Particular BVP problem is defined by:

(1) numbers s_0 and s_1 specifying boundary;

(2) functions $c(x), b(x), a(x)$ and $f(x)$ specifying BVP coefficients;

(3) type of boundary conditions at the points s_0 and s_1.

There is nothing special about item (1) while representation of (2) and (3) in terms of MATLAB structures needs to be clarified. Functions in MATLAB can be defined either through anonymous inline declaration

```
f = @(x) sin(x)
```

or using stand-along .m file containing function definition

```
function y=f(x)
  y=sin(x);
end
```

Whenever it is possible, we strive to use vectorized computations which are much more efficient in terms of computational costs. Having a vector

```
x = [x1,...,xn],
```

function call `y = f(x)` returns a vector

```
y = [f(x1),...,f(xn)]
```

and we operate with variables being vectors rather than with individual components.

Considering item (3), we can use matrix to specify boundary conditions. Let us introduce a matrix bc having dimension 2×3 which packs all boundary condition data. First row corresponds to BC at the point s0 and the second row contains data for the boundary point s1. First column indicates type of BC while second and third columns contain corresponding data values. For example, bc(1,:)=[1,0,u_s0] is for the BC of the first type at the point s0 and bc(2,:)=[3,sigma,mu] is for the BC of the third type at the point s1.

Below, MATLAB codes for algorithm implementation will be given in reverse order with regards to above considered BEM algorithmic steps, namely, we start with coding quadrature rules and finish with solving linear system of the BEM. All implementations will be based on derived expressions.

2.3 CODING QUADRATURE RULES

Nodes and weights of Gaussian quadrature on arbitrary interval [s1,s2] can be calculated using listing 2.1. Function GaussQuad() inputs the number of points, n, left boundary of the interval, s0, right boundary, s1, and returns a pair of vectors, d and w, which are quadrature nodes and weights accordingly. Here, quadrature nodes are zeroes of Legendre polynomial. The function can take either three or one argument and in case of a single argument, it is assumed the nodes are sought on the interval [−1,1]. This logic is implemented in lines 2–4 with MATLAB function nargin which returns the number of function input arguments given in the call to the currently executing function. The code lines 8–10 compute the elements of matrix Q (see (1.87)) and compose the matrix using diagonal assembling. Function diag(bta,k) places the elements of vector bta on the k-th diagonal (k=0 represents the main diagonal, k=1 is above the main diagonal, and k=−1 is below the main diagonal). Then, we perform eigen value decomposition (line 11) and extract sorted eigen values (line 12). To perform eigen values extraction, first, we use MATLAB function diag() which being called with matrix L returns its main diagonal, then we call the function sort() which returns sorted nodes, d, and it also returns an index vector, k, that describes the rearrangement of elements such that d = L(k,:). We need this index vector to keep correspondence between sorted eigen values and unsorted eigen vectors, U. The last step is to calculate weights on the interval [−1,1] (line 13) and to perform transformation for extended region [s0,s1] (lines 14–15). Note that all calculations here are vectorized, namely, we do not use loops to iterate over vector or matrix elements, but rather use vectors itself in calculation formulas. This significantly improves performance and simplifies code readability.

```
1 function [d,w] = GaussQuad(n,s0,s1)
2   if nargin==1
3      s0=-1; s1=1;
4   end
5   if n==1
6      d=(s0+s1)/2; w=s1-s0; return;
7   end
8   j=1:n-1; bta=j./sqrt(4*j.^2-1); Q=diag(bta,-1)+diag(bta,1);
9   [U,L]=eig(Q);
10  [d,k]=sort(diag(L));
11  w=2*U(1,k).^2; w=w(:);
12  d=s0+0.5*(s1-s0)*(d+1);
13  w=0.5*(s1-s0)*w;
14 end
```

Listing 2.1: MATLAB code for Gaussian quadrature rule.

Similarly, Lobatto quadrature implementation can be coded as shown in listing 2.2. We need to make sure that input number of nodes is equal or greater than two to make calculation possible (lines 2–5). Compared with Gaussian quadrature rule, matrix elements are calculated differently (line 14) and boundary points are added to the nodes and weights (lines 19–20). Recall that the nodes of Lobatto quadrature can be taken to be interpolation nodes on the interval [s0,s1].

```
1 function [d,w] = LobattoQuad (n,s0,s1)
2   if n<2
3     disp ('LobattoQuad: number of nodes is less then 2' );
4     return;
5   end
6   if nargin==1
7     s0=-1; s1=1;
8   end
9   if n==2
10    d=[s0;s1]; w=[(s1-s0)/2; (s1-s0)/2];
11    return;
12  end
13  j=1:n-3;
14  bta=sqrt(j.*(j+2)./(2*j+1)./(2*j+3));
15  Q=diag(bta,-1)+diag(bta,1);
16  [U,L]=eig(Q);
17  [d,k]=sort(diag(L));
18  w=4/3*(U(1,k).^2)';
19  w=[2/(n^2-n); w./(1-d.^2); 2/(n^2-n)];
20  d=[-1; d ; 1 ];
21  d=s0+0.5*(s1-s0)*(d+1);
22  w=0.5*(s1-s0)*w;
23 end
```

Listing 2.2: MATLAB code for Lobatto quadrature rule.

2.4 IMPLEMENTATION CODE FOR INTERPOLATING AND DIFFERENTIATING MATRICES

Interpolating and differentiating matrices ($\hat{E}$ and $\hat{D}$) are used for calculation of the function u_h and its derivative at arbitrary points. These matrices are defined by basis functions for calculation of which we utilize expressions (1.13) and (1.14)

$$\beta_\alpha = \left(\prod_{j=1, j \neq \alpha}^{m+1} (\hat{x}_\alpha - \hat{x}_j) \right)^{-1}, \tag{2.1}$$

$$\hat{E}_{k,\alpha} = \hat{\varphi}_\alpha(\hat{d}_k) = \beta_\alpha \prod_{j=1, j \neq \alpha}^{m+1} (\hat{d}_k - \hat{x}_j), \tag{2.2}$$

$$\hat{D}_{k,\alpha} = \hat{\varphi}'_\alpha(\hat{d}_k) = \beta_\alpha \sum_{l=1, l \neq \alpha}^{m+1} \prod_{j=1, j \neq \alpha, j \neq l}^{m+1} (\hat{d}_k - \hat{x}_j). \tag{2.3}$$

The code in listing 2.3 implements calculation of $\hat{E}$ and $\hat{D}$ defined on the basis element. Function interpolatingMat() returns interpolation matrix, E, and differentiation matrix, D, with input arguments to be interpolation points, x, and arbitrary nodes at which interpolation is calculated, d. Since the formulas for matrix elements are nonlinear we cannot fully apply vectorized calculations and have to use loops to iterate over elements. First, elements β_α are calculated in the loop lines 4–7 where the variable b is used to store the values of β_α. Vector j forms indices which span the range of product operator in (2.1) and function prod() returns the product of the vector elements. Listing lines 8–21 implement matrices E and D calculations with two nested loops for matrix elements k and i to calculate the product and one more loop to calculate the sum in (2.3). The indices are processed consistently in ascending order, so lines 14–15 perform this logic for product and summation calculation. Indices formation in the lines 5, 9 and 15 follows the same pattern with skipping i-th element.

```
1  function [E,D]=interpolatingMat(x,d)
2      nx=numel(x); nd=numel(d);
3      E=zeros(nd,nx); D=zeros(nd,nx); b=zeros(1,nx);
4      for i=1:nx
5          j=[1:(i-1),(i+1):nx];
6          b(i)=1/prod(x(i)-x(j));
7      end
8      for i=1:nx
9          j=[1:(i-1),(i+1):nx];
10         for k=1:nd
11             E(k,i)=b(i)*prod(d(k)-x(j));
12             ds=0;
```

```
13       for l=j
14         i1=min(i,l); i2=max(i,l);
15         jj=[1:(i1−1),(i1+1):(i2−1),(i2+1):nx];
16         ds=ds+prod(d(k)−x(jj));
17       end
18       D(k,i)=b(i)*ds;
19     end
20   end
21 end
```

Listing 2.3: MATLAB code for interpolating and differentiating matrices calculation.

In addition to the function interpolatingMat(), it is useful to have a functionality which allows performing mesh refinement and calculating function values along with its derivatives on the refined mesh. This is helpful, for example, when plotting solution. Listing 2.4 illustrates the code for generating corresponding refined mesh values. It makes use previous function interpolatingMat() to obtain matrices E and D and new mesh is generated by splitting each element into k subelements. Function getRefinedValues() has following arguments: u is a vector of function values on the coarse mesh, xl is a vector of finite element nodes, d is a vector of points on the basis element [0,1] and k is a number of refined mesh points on each finite element. Returning vectors are refined mesh points, x, new function values, y, and new function derivative values on refined mesh, dy. New points are generated using MATLAB command linspace(x1,x2,k) which generates k evenly distributed points in the interval [x1,x2] (see lines 2 and 12). For better readability, listing 2.4 is complemented with comments.

```
1 function [x,y,dy]=getRefinedValues(u,xl,d,k)
2    d_ref=linspace(0,1,k+1); % refined basis element
3    [E,D]=interpolatingMat(d,d_ref);
4    n=numel(xl); m=numel(d)−1;
5    x_ref=(n−1)*k+1; % number of all refined mesh points
6    x=zeros(x_ref,1); y=zeros(size(x)); dy=zeros(size(x));
7    % Iterate over finite elements and refine them
8    for l=1:n−1
9      h=xl(l+1)−xl(l); % size of element l
10     i_c=(l−1)*m+(1:m+1); % coarse point indices
11     i_ref=(l−1)*k+(1:k+1); % refined points indices
12     x(i_ref)=linspace(xl(l),xl(l+1),k+1)';
```

```
13    y(i_ref)=E*u(i_c);
14    dy(i_ref)=D*u(i_c)/h;
15  end
16 end
```

Listing 2.4: MATLAB code for generating values on refined mesh.

So far, we have created flexible computational tools for manipulation with mesh and quadrature points which lays a foundation for the next step to implement numerical integration.

EXERCISES

▶ Verify that the number of operations to calculate E and D using function interpolatingMat() has the order of $O(\mathtt{nx}^2\mathtt{nd})$ and $O(\mathtt{nx}^3\mathtt{nd})$, respectively.

▶ Perform a number of running function interpolatingMat() with different value of dimension nx when x=d and estimate running time. Define dependence of calculation time on value nx.

▶ Show that when x=d then the matrix D can be calculated through

$$\hat{D}_{ij} = \frac{\beta_j}{\beta_i(\hat{x}_i - \hat{x}_j)}, \quad i \neq j, \quad \hat{D}_{ij} = -\sum_{j=1,j\neq i}^{n} \hat{D}_{ij}.$$

Usage of these expressions results in complexity of calculation matrix D to be $O(\mathtt{nx}^2)$ rather than $O(\mathtt{nx}^4)$ compared to the implementation function interpolatingMat().

▶ Compose your own version of function interpolatingMat().

2.5 IMPLEMENTATION CODE FOR CALCULATING AND ASSEM-BLING FEM MATRICES

As discussed above, composing of the global stiffness matrix, $\tilde{A}$, and global forcing vector, $\tilde{F}$, can be decomposed into two parts (see 1.73 and 1.77): assembling matrix A and vector F which do not take into account boundary condition and assembling matrix S and vector M which account for boundary conditions. Function assemblingAF() in linsting 2.5 implements the first part where we compute A and F.

Generally speaking, the function assemblingAF() returns stiffness matrix A and forcing vector F corresponding to the equation $-(cu')' + bu' +$

$au = f$ with homogeneous Neumann boundary conditions $cu' = 0$. Arguments have the following meaning: c,b,a,f are equation coefficients, xl is a vector of finite elements nodes, x is a vector of interpolation points on the basis element, d is a vector of quadrature nodes on the basis element and w is a vector of quadrature weights on the basis element. The code fully implements algorithm 3 for calculating elements of local matrices and then it incorporates assembling global matrices without taking into account boundary conditions (first two steps of algorithm 2). Code line 5 corresponds to the first step of algorithm 2 and lines 6–17 implement second step with looping over finite elements, computing local matrices (lines 7–14) and accumulating components of global matrices (lines 15–16).

```
1  function [A,F]=assemblingAF(c,b,a,f,xl,x,d,w)
2    [E,D]=interpolatingMat(x,d);
3    ET=E'; DT=D';
4    m=numel(x)−1; n=numel(xl); n_total=(n−1)*m+1;
5    A=sparse(n_total,n_total); F=zeros(n_total,1);
6    for l=1:n−1
7      h=xl(l+1)−xl(l); % size of element l
8      ne=(l−1)*m+(1:m+1); % point indices
9      xd=xl(l)+h*d; % quadrature nodes on element l
10     c_loc=diag((w/h).*c(xd));
11     b_loc=diag(w.*b(xd));
12     a_loc=diag((h*w).*a(xd));
13     Al=(DT*c_loc+ET*b_loc)*D+ET*a_loc*E;
14     Fl=(h*w).*f(xd);
15     A(ne,ne)=A(ne,ne)+Al;
16     F(ne)=F(ne)+ET*Fl(:);
17   end
18 end
```

Listing 2.5: MATLAB code for assembling one-dimensional matrices.

It is worth stressing attention on the line A(ne,ne)=A(ne,ne)+Al. Here, A is n_total by n_total global stiffness matrix, Al is (m+1) by (m+1) local stiffness matrix and the structure ne contains global indices of points on each element. As such, local element is added to the corresponding component of the global matrix which represents assembling the global matrix. This assembling strategy is not fully vectorized but it is good for matrix of moderate number of components. Later, in section 5.1, we consider other approaches to assemble matrices which are efficient for large matrices and allow full vectorization.

Next, we need to take into account BC and calculate matrix S and vector M which are being added to A and F, respectively, to form full FEM matrices. Incorporating BC is performed according to the step three of algorithm 1 for BC of the first type (Dirichlet BC), and according to the step three of algorithm 2 for BC of the third type (Robin BC). Recall that BC of the second type (Neumann BC) can be considered as a special case of Robin BC. Code listing 2.6 implements logic for including BC where function assemblingBC() inputs boundary condition matrix, bc, stiffness matrix, A, forcing vector, F, and outputs stiffness matrix and forcing vector complemented with BC, [A0,F0]. As discussed in section 2.2, first column of matrix bc encodes type of BC, so lines 6–13 correspond to BC of the third type either for point s0 (lines 6–9) or point s1 (lines 10–13). Then, BC of the first type is considered (lines 15–22). In both cases, the stiffness matrix and forcing vector are modified in some way to include BC. The last step is to form final matrices from the modified one (lines 23–25).

```
1  function [A0,F0]=assemblingBC(bc,A,F)
2  %  .bc - boundary conditions matrix of size 2x3 :
3  %    bc(1,:)=[1,0,u_s0] , or bc(1,:)=[3,sig_s0,mu_s0]
4  %    bc(2,:)=[1,0,u_s1] , or bc(2,:)=[3,sig_s1,mu_s1]
5    n=numel(F);% number of all mesh points
6    if bc(1,1)==3
7       A(1,1)=A(1,1)+bc(1,2);
8       F(1)=F(1)+bc(1,3);
9    end
10   if bc(2,1)==3
11      A(n,n)=A(n,n)+bc(2,2);
12      F(n)=F(n)+bc(2,3);
13   end
14   ib=1; ie=n;
15   if bc(1,1)==1
16      ib=2; u_s0=bc(1,3);
17      F=F—u_s0*A(:,1);
18   end
19   if bc(2,1)==1
20      ie=n—1; u_s1=bc(2,3);
21      F=F—u_s1*A(:,n);
```

```
22    end
23    I=(ib:ie)';
24    A0=A(I,I);
25    F0=F(I);
26 end
```

Listing 2.6: MATLAB code for assembling one-dimensional boundary conditions.

Once the final stiffness matrix, A0, and forcing vector, F0, are created using function assemblingBC(), resulting FEM system can be solved and solution can be composed according to the steps 4 and 5 of algorithm 1. Function solveFEM() in listing 2.7 returns the final solution, u. The code lines 12–13 call linear solver (backslash MATLAB operator) and append Dirichlet BC if any (u_s0 and u_s1).

```
1 function u = solveFEM(bc,A0,F0)
2    u_s0 =[]; u_s1=[];
3    if bc(1,1)==1 , u_s0=bc(1,3) ; end
4    if bc(2,1)==1 , u_s1=bc(2,3) ; end
5    u0=A0\F0;
6    u=[u_s0;u0;u_s1];
7 end
```

Listing 2.7: MATLAB code for solving one-dimensional FEM system.

It is convenient to combine previous three functions into the single one which performs solution of a linear BVP. Let us introduce a function solveBVP() which return final solution, u, and mesh points, x_mesh, with given arguments being BC, equation coefficients, finite element nodes, and interpolation along with quadrature nodes on the basis element (see listing 2.8).

```
1 function [u,x_mesh]=solveBVP(bc,c,b,a,f,xl,x,d,w)
2    m=numel(x)−1; n=numel(xl);
3    N=(n−1)*m+1; % number of mesh points
4    x_mesh=zeros(N,1); % mesh points
5    for l=1:n−1
6       ne=(l−1)*m+(1:m+1); % mesh point indices
7       x_mesh(ne)=xl(l)+(xl(l+1)−xl(l))*x;
8    end
9    [A,F]=assemblingAF(c,b,a,f,xl,x,d,w);
10   [A0,F0]=assemblingBC(bc,A,F);
```

```
11    u=solveFEM(bc,A0,F0);
12 end
```

Listing 2.8: MATLAB code for solving one-dimensional BVP.

We could stop here and decide that the task of coding FEM is done, but one important part of any numerical simulation pipeline is missing. To make sure that we have done all correctly and to validate output results, we need a tool for error analysis. The main approach is to run simulation for a problem with known exact solution and then compare the output with true values in the form of error. To analyse FEM error when exact solution is known, it is worth having a function which calculates the error norm in spaces L_2 and H^1

$$e = u - u_h, \quad \|e\|_{L_2}^2 = \int_{s_0}^{s_1} e^2(x)dx, \quad \|e\|_{H^1}^2 = \int_{s_0}^{s_1} \left(e'(x)\right)^2 dx.$$

Function errorNorm() which code is given in listing 2.9 calculates and outputs these norms. Here, e0 is a value of L_2 error norm, e1 is a value of H^1 error norm, u and du are exact solution and its derivative, y is an FEM solution, xl is a vector of finite element nodes and d is a vector of interpolation points on the basis element [0,1]. Function errorNorm() uses Gaussian quadrature to calculate the norms, and interpolating and differentiating matrices to find corresponding values of the solution.

```
1 function [e0,e1]=errorNorm(u,du,y,xl,d)
2    m=numel(d)—1;
3    [d_new,w]=GaussQuad(m+2,0,1);
4    [E,D]=interpolatingMat(d,d_new);
5    e0=0; e1=0;
6    for l=1:numel(xl)—1
7       h=xl(l+1)—xl(l);
8       dl=xl(l)+h*d_new;
9       ic=(l—1)*m+(1:m+1);
10      uh=E*y(ic); duh=D*y(ic)/h;
11      e=u(dl)—uh; de=du(dl)—duh;
12      e0=e0+h/2*sum(w.*e.^2); e1=e1+h/2*sum(w.*de.^2);
13   end
14   e0=e0^(1/2); e1=e1^(1/2);
15 end
```

Listing 2.9: MATLAB code for the FEM error analysis.

Now, we are ready to solve example problem using created library. Let us consider the following BVP

$$(u'\cos(x))' - 2u'\sin(x) + u\sin(x) = \sin^2(x), \quad x \in (-1,1),$$
$$u(-1) = \sin(-1), \quad u'(1) = \cos(1). \quad (2.4)$$

Here, exact solution to be $u = \sin(x)$ and equation coefficients are the following functions : $c(x) = \cos(x)$, $b(x) = -2\sin(x)$, $a(x) = \sin(x)$, $f(x) = \sin^2(x)$. Boundary points are $s_0 = -1$ and $s_1 = 1$. At the point s_0 we have Dirichlet BC which is a BC of the first type and at the point s_1 we have Neumann BC. The FEM library deals with either first or third type of BC, so we need to rewrite BC at the point s_1 in the form of BC of the third type, namely, in the form $c(1)u'(1) + \sigma(1)u(1) = \mu(1)$. It is easy to see that $\sigma = 0$ and $\mu = \cos^2(1)$.

Having all basic functions for assembling and solving FEM, it is not difficult to write a script (see listing 2.10) which solves example BVP and plots solution along with error norm values.

```
1 % main script
2 clc ; clear all ; close all
3 % set BVP data:
4 % integration domain
5 s_0=-1; s_1=1;
6 % exact solution
7 u=@(x) sin(x);
8 du=@(x) cos(x);
9 % equation coefficients
10 c=@(x) cos(x);
11 b=@(x) -2*sin(x);
12 a=@(x) sin(x);
13 f=@(x) sin(x).^2;
14 % boundary conditions
15 sigma=0;
16 mu=c(s_1)*du(s_1)+sigma*u(s_1);
17 bc=[1 0 u(s_0)
18     3 sigma mu ];
19 % set FEM parameters
20 m=3; % polynomial degree
21 ne=6; % number of final elements
```

```
22 M=m+1; % number of quadrature nodes
23 xl=linspace(s_0,s_1,ne+1); % finite element nodes
24 t=LobattoQuad(m+1,0,1); % interpolation nodes on basis element
25 [d,w]=LobattoQuad(M,0,1); % quadrature nodes and weighs
26 % solve BVP
27 tic
28 [uh,x]=solveBVP(bc,c,b,a,f,xl,t,d,w);
29 timeFemSol=toc
30 % find error in various norms
31 [e0,e1]=errorNorm(u,du,uh,xl,t)
32 [X,Uh,dUh]=getRefinedValues(uh,xl,t,10*m);
33 Z=u(X)—Uh; z=u(x)—uh; dZ=du(X)—dUh;
34 ie=1:m:numel(x);
35 einfh=norm(Z,inf)
36 einfdh=norm(dZ,inf)
37 einfbh=norm(u(xl(:))—uh(ie),inf)
38 figure; % plot solution
39 plot(X,Uh,'—b',x,uh,'rx',x(ie),uh(ie),'ro')
40 xlabel('x'); ylabel('u_h')
41 figure;% plot solution error
42 plot(X,Z,'—b',x,z,'rx',x(ie),z(ie),'ro')
43 xlabel('x'); ylabel('u—u_h')
44 figure;% plot derivatives error
45 plot(X,dZ,'—b',x,zeros(size(x)),'rx',x(ie),zeros(size(x(ie))),'ro')
46 xlabel('x'); ylabel('u'—u''_h')
```

Listing 2.10: MATLAB code for solving example one-dimensional BVP.

Plots generated by this script are shown in figures 2.1, 2.2 and 2.3 where the finite element nodes are indicated as circles and crosses are quadrature nodes which coincide with interpolation nodes.

Estimations of numerical error for the FEM solution are given in table 2.1. It can be seen that implementation of the FEM algorithm performs numerical calculations with acceptable accuracy.

TABLE 2.1: Summary of computational errors for the FEM solution to the problem (2.4).

$\|e\|_{L_2}$	$\|e\|_{H^1}$	$\|e\|_\infty$	$\|e'\|_\infty$
10^{-6}	10^{-5}	10^{-6}	10^{-4}

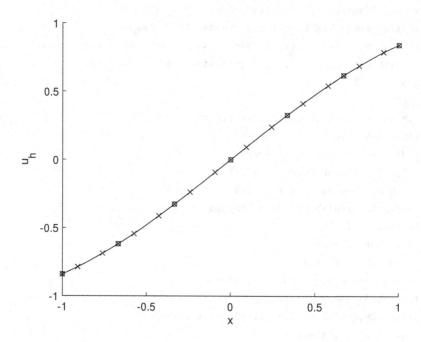

FIGURE 2.1: FEM solution to the problem (2.4).

Changing equation coefficients and parameters of the FEM algorithm in the main script allows solving various BVP. A reader is encouraged to play around with different parameter settings (m,ne,M) and see how it affects solving time and errors magnitude.

EXERCISES

▶ Find numerical solution to the problems given below with known exact solutions. Boundary conditions of the first type may be chosen to close BVP. Specify your own values for parameter ε. Perform analysis of the solution behaviour when ε becomes small. Choose different FEM parameters and make observation regarding changes

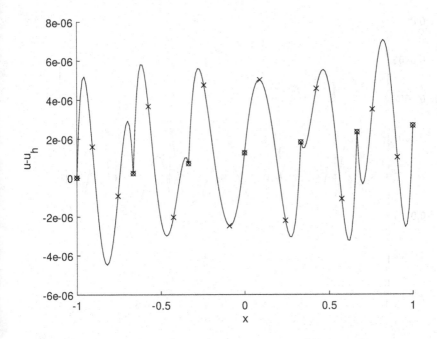

FIGURE 2.2: Absolute error of the FEM solution to the problem (2.4).

of the error magnitude.

1. $-\varepsilon u'' + u = 0, \quad x \in (0,1), \quad u = \dfrac{\exp\left(-\frac{x}{\sqrt{\varepsilon}}\right) - \exp\left(\frac{x-2}{\sqrt{\varepsilon}}\right)}{1 - \exp\left(-\frac{2}{\sqrt{\varepsilon}}\right)}.$

2. $-\varepsilon u'' - bu' + u = (1 + \varepsilon\pi^2)\cos(\pi x) + b\pi\sin(\pi x), \quad x \in (-1,1),$
   $b = 2 + \cos(\pi x), \quad u = \cos(\pi x).$

3. $-(\varepsilon + x^2)u'' - 4xu' - 2u = 0, \quad x \in (1,1), \quad u = \dfrac{1}{\varepsilon + x^2}.$

4. $-\varepsilon u'' + u = (1 + \varepsilon\pi^2)\cos(\pi x), \quad x \in (-1,1),$
   $u = \cos(\pi x) + \exp\left(\dfrac{x-1}{\sqrt{\varepsilon}}\right) + \exp\left(-\dfrac{x+1}{\sqrt{\varepsilon}}\right).$

2.6 PYTHON IMPLEMENTATION

To give a reader more flexibility in implementing FEM, we turn our attention to the Python programming language and explain how to create

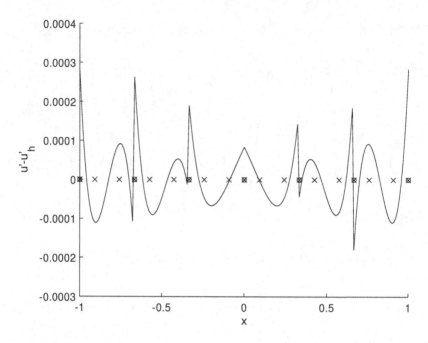

FIGURE 2.3: Absolute error of the FEM solution derivative corresponding to the problem (2.4).

FEM computational framework in Python. Basically, we are going to translate MATLAB codes into Python ones with some variations in specific features. We expect that a reader has basic Python programming skills.

For dense matrix operation we use NumPy library which provides broad range of specific functions for matrix manipulation and most of them have MATLAB analogy. Book [7] could be helpful in improving scientific programming skills using NumPy library.

There are some important notes regarding using vectors and matrices in Python. First, it is worth noting that indexing in Python starts with zero (unlike indexing in MATLAB starting with one) and all index values given in previous chapters should be decremented by one to correspond Python indexing.

Second, there is a distinction between different types of array data structure in Python. For example, there is a structure called list, a=[1,2,3], which stores data in array like manner, but it is not a vector in terms of ability to be used in matrix calculations; however if it is declared through NumPy library, `a = numpy.array([1, 2, 3])`, then it can be used in vectorized calculations, for example, in a dot product calculation of two arrays, `numpy.dot(a,b)`. Conversion between list data structure to the `numpy.array` structure is supported by the NumPy library. Another note is that unlike in MATLAB where most built in functions support vectorized calculations using any array variable, in Python these vectorized calculations should be performed through explicitly calling NumPy library commands. For example, MATLAB version of square root function call on vector, `j=1:n;s=sqrt(j.)`, in Python would have the form: `j=numpy.arange(1,n);s=numpy.sqrt(j)`.

Also, unlike in many matrix languages, the product operator, `*`, operates elementwise in NumPy arrays, and matrix product can be performed using the `numpy.matmul()` or `@` operator (in Python version >=3.5). For example, having two matrices `A=numpy.array([[1, 0],[0, 1]])` and `B=numpy.array([[4, 1],[2, 2]])`, result of operation `A*B` would be `array([[4, 0],[0, 2]])`, but `A@B` results in `array([[4, 1],[2, 2]])`.

Array slicing and indexing works exactly the same way it does for lists except that they can be applied to multiple dimensions as well. For example, given a matrix `A`, the command `A[0,3]` accesses a single element of the matrix, and `A[:,3]` extracts a column. Note that slices of arrays do not copy the internal array data but only produce new views of the original data, so if `b=A[0,3]` and `b=b+1` then element of matrix `A` would be modified implicitly as well. If a copy is needed, then an explicit `copy()` command is required. NumPy arrays may be indexed with other arrays or lists. For all cases of index arrays, the copy of original data is returned, not a view. For example `A[np.array([0,2,4]),0]` or `A[np.array([0,2,4])][0]` returns elements of the first column with indices 0, 2 and 4.

To operate with sparse matrices in Python we use `scipy.sparse` library. There are various sparse matrix classes in `scipy.sparse` library (`coo_matrix`, `csc_matrix`, `csr_matrix`, etc.) to perform matrix manipulations of different types. For example, class `coo_matrix` is used for constructing a sparse matrix in coordinate format from arrays of indices and values (see figure 2.1 for detailed explanations) but it does not directly support arithmetic operations. Class `sparse.lil_matrix` is a structure for constructing sparse matrices incrementally which supports flexible slicing. Classes `csr_matrix` and `csc_matrix` are for compressed sparse row matrix and compressed sparse column matrix representations, respectively. These matrix formats are suitable for arithmetic operations and row/column slicing but inefficient for changing sparsity structure. Usually, `coo_matrix` is a fast format for constructing sparse matrices. Once a matrix has been constructed, conversion to `csc_matrix` or `csr_matrix` format is performed (explicitly or implicitly) for fast arithmetic and matrix vector operations.

Before start writing Python functions, it is required to import libraries which we will be using across the code. Script in listing 2.11 shows how to perform importing. Libraries `numpy` and `scipy.sparse` are for working with dense and sparse matrices, `spsolve` is for solving sparse linear systems, `plt` is for plotting simulation results and `timeit` library allows operate with timing counters. We need to run this script only once in the FEM Python library.

```
1 import numpy
2 import scipy.sparse as sparse
3 import timeit
4 from scipy.sparse.linalg import spsolve
5 import matplotlib.pyplot as plt
```

Listing 2.11: Python code implementing libraries import.

Python listings given below contain functions which names and functionality repeat ones written in MATLAB, so notations of arguments and returning values have consistent with MATLAB counterparts meaning. While providing Python implementation of particular function, the focus will be on specific Python language features which differ the code from MATLAB version rather than on describing logic, which already has been discussed in previous sections.

Let us look into Python implementation of Gaussian and Lobatto quadrature rules (see listings 2.12 and 2.13). By design, arguments `s0` and `s1` can be omitted, so the code line 2 retrieves the number of actual input arguments and stores it in the variable **nargin**. This is done through the function `locals()` which returns a vector of arguments, and the function `len()` returning vector length.

We can see in the code arithmetic calculation of the form `j**2`, which implements raising `j` to the second power. The function `numpy.diag(bta,k)` constructs a diagonal matrix from a given array, `bta`, and a diagonal position, `k`. The function `linalg.eig(Q)` computes eigenvalues and eigenvectors of the matrix `Q`. To obtain ordered eigenvalues, function `numpy.argsort(L)` is used which returns the indices that would sort the array `L`.

```
1 def GaussQuad(n,s0,s1):
2     nargin=len(locals())
3     d=[]; w=[]
4     if nargin==1:
```

```
5          s0=-1; s1=1
6      if n==1:
7          d=(s0+s1)/2; w=s1-s0
8          return d,w
9      j=numpy.arange(1,n)
10     bta=j/numpy.sqrt(4*(j**2)-1)
11     Q=numpy.diag(bta, k=-1)+numpy.diag(bta, k=1)
12     L,U = numpy.linalg.eig(Q)
13     k=numpy.argsort(L)
14     d=L[k]; w=2*(U[0][k]**2)
15     d=s0+0.5*(s1-s0)*(d+1)
16     w=0.5*(s1-s0)*w
17     return d,w
```

Listing 2.12: Python code for Gaussian quadrature rule.

Code lines 18–21 in listing 2.13 implement extending arrays of weights, w, and nodes, d, with new values at the boundary. This is coded through the functions numpy.append() to append value to the end of the array, and numpy.insert() to insert the value at the beginning of the array. Overall, calculations follow linear logic of the corresponding algorithms for Gaussian and Lobatto quadrature rules.

```
1  def LobattoQuad (n,s0,s1):
2      nargin=len(locals())
3      d=[]; w=[]
4      if n<2:
5          print('LobattoQuad: number of nodes is less then 2')
6          return d,w
7      if nargin==1:
8          s0=-1; s1=1
9      if n==2:
10         d=[s0,s1]; w=[(s1-s0)/2,(s1-s0)/2]
11         return d,w
12     k=numpy.arange(1,n-2)
13     bta=numpy.sqrt(k*(k+2)/(2*k+1)/(2*k+3))
14     Q=numpy.diag(bta,k=-1)+numpy.diag(bta,k=1)
15     L,U = numpy.linalg.eig(Q)
16     k=numpy.argsort(L)
17     d=L[k]; w=4/3*(U[0][k]**2); w=w/(1-d**2)
18     w=numpy.append(w,2/(n**2-n))
19     w=numpy.insert(w,0,2/(n**2-n))
```

```
20      d=numpy.append(d,1)
21      d=numpy.insert(d,0,-1)
22      d=s0+0.5*(s1-s0)*(d+1);
23      w=0.5*(s1-s0)*w;
24      return d,w
```

Listing 2.13: Python code for Lobatto quadrature rule.

Script listing 2.14 demonstrates how interpolating and differentiating matrices calculation can be implemented in Python. Note that it is a good practice to declare and initialize variables before using them to make it clear of what data manipulations are allowed to work with. Lines 2–6 introduce main variables which are used across the listing where nx and nd represent lengths of corresponding input arrays, E and D are declared to be zeroth matrices of dimension nd by nx, and b is a one-dimensional array of length nx. Function numpy.**zeros**((nd,nx)) returns a matrix of zeros with the given shape, (nd,nx), and if the second dimension is omitted, then a vector is returned. Looping with parameter can be coded through either specifying a range using a special function range(nx) (see line 7) which returns iterator spanning values in the range [0,nx), or giving the range explicitly as a vector of predefined values (see line 15). Formulas (2.1)–(2.3) which are coded in listing 2.14 include calculation of product of elements which is implemented through function numpy.**prod**(). Indices over which the product in formulas is taken exclude particular values and this is implemented using numpy.**delete**() function which returns a new array with deleted sub-arrays (lines 8, 11 and 17).

```
1 def interpolatingMat(x,d):
2      nx=numpy.size(x); nd=numpy.size(d)
3      E=numpy.zeros((nd,nx))
4      D=numpy.zeros((nd,nx))
5      b=numpy.zeros((nx,))
6      j_all=numpy.arange(nx)
7      for i in range(nx):
8          j=numpy.delete(j_all,i)
9          b[i]=1/numpy.prod(x[i]-x[j])
10     for i in range(nx):
11         j=numpy.delete(j_all,i)
12         for k in range(nd):
13             E[k,i]=b[i]*numpy.prod(d[k]-x[j])
14             ds=0
15             for l in j:
```

```
16                      i1=min(i,1); i2=max(i,1)
17                      jj=numpy.delete(j_all,[i1,i2])
18                      ds=ds+numpy.prod(d[k]—x[jj])
19                  D[k,i]=b[i]*ds
20      return E,D
```

Listing 2.14: Python code for interpolating and differentiating matrices calculation.

Along with calculation of interpolating and differentiating matrices, error norms calculations and mesh refinement are another complementary functionalities which simplify manipulating FEM data (see listings 2.15 and 2.16). Note that matrix multiplication is performed using @ operator.

```
1  def errorNorm(u,du,y,xl,d):
2      m=numpy.size(d)—1
3      d_new,w=GaussQuad(m+2,0,1)
4      E,D=interpolatingMat(d,d_new)
5      e0=0; e1=0
6      for l in range(numpy.size(xl)—1):
7          h=xl[l+1]—xl[l]
8          dl=xl[l]+h*d_new
9          ic=l*m+numpy.arange(m+1)
10         uh=E @ y[ic]; duh=D @ y[ic]/h
11         e=u(dl)—uh; de=du(dl)—duh
12         e0=e0+h/2*sum(w*(e**2))
13         e1=e1+h/2*sum(w*(de**2))
14     e0=e0**(1/2); e1=e1**(1/2)
15     return e0,e1
```

Listing 2.15: Python code for the FEM error analysis.

```
1  def getRefinedValues(u,xl,d,k):
2      d_ref=numpy.linspace(0,1,k+1) # refined basis element
3      E,D=interpolatingMat(d,d_ref)
4      n=numpy.size(xl); m=numpy.size(d)—1
5      x_ref=(n—1)*k+1; # number of all refined mesh points
6      x=numpy.zeros((x_ref,))
7      y=numpy.zeros(x.shape)
8      dy=numpy.zeros(x.shape)
9      for l in range(n—1):
```

```
10        h=xl[l+1]−xl[l]  #size of element l
11        i_c=l*m+numpy.arange(m+1) # coarse point indices
12        i_ref=l*k+numpy.arange(k+1)  # refined points indices
13        x[i_ref]=numpy.linspace(xl[l],xl[l+1],k+1)
14        y[i_ref]=E @ u[i_c]
15        dy[i_ref]=D @ u[i_c]/h
16    return x,y,dy
```

Listing 2.16: Python code for generating values on refined mesh.

Let us discuss in more details listing 2.17 for assembling FEM matrices. Here, we use sparse matrices as a main data structure for stiffness matrix, A, and forcing vector, F, and declare them through sparse.coo_matrix() class. This class allows creating matrices in coordinate form using constructor coo_matrix((data, (i, j)), shape=(m, n)), where (m,n) are the matrix dimensions, data is an array of matrix entries, i is an array of row indices and j is an array of column indices of the matrix entries. If arguments (data, (i, j)) are omitted, then an empty sparse matrix of given shape will be created, as shown in code lines 8 and 9. Class for sparse matrices in Python does not support advanced indexing, so the best way of assembling stiffness matrix is to use extended local matrix as a summand when accumulating final stiffness matrix (see definition 1.10 and algorithm 2). Recall that extended local stiffness matrix has all zeros apart from the diagonal block for corresponding final element. Assume that we have a final element with three nodes, so the block would be $3 \times 3 = 9$ elements and to construct a bock diagonal sparse matrix we need three arrays with nine elements each, namely, indices i=[0,0,0,1,1,1,2,2,2], j=[0,1,2,0,1,2,0,1,2] and data with some data values which are flatten one-dimensional representation of two-dimensional (3×3) block. Generating local indices is implemented in lines 10–12, where i_loc is a range of local node indices, i_loc_sp_mat is an array of local row indices and j_loc_sp_mat is an array of local column indices for the sparse matrix constructor. Here, function numpy.repeat(i_loc,m+1) repeats each element of the array i_loc given number of times, m+1, and numpy.tile(i_loc,m+1) construct an array by repeating entire array i_loc the number of times, m+1. After calculating local matrix in dense form (line 19), global row and column indices are calculated (lines 20–21), and extended stiffness matrix is added in the form of sparse matrix (see line 22). This approach illustrates a possible way of manipulating with sparse matrices.

```
1 def assemblingAF(c,b,a,f,xl,x,d,w):
```

```
2       E,D=interpolatingMat(x,d)
3       ET=numpy.transpose(E)
4       DT=numpy.transpose(D)
5       m=numpy.size(x)—1
6       n=numpy.size(xl)—1
7       n_total=n*m+1;
8       A=sparse.coo_matrix((n_total,n_total))
9       F=sparse.coo_matrix((n_total,1))
10      i_loc=numpy.arange(m + 1)
11      i_loc_sp_mat = numpy.repeat(i_loc, m + 1)
12      j_loc_sp_mat = numpy.tile(i_loc, m + 1)
13      for l in range(n):
14          h=xl[l+1]—xl[l] # size of element l
15          xd=xl[l]+h*d # quadrature nodes on element l
16          c_loc=numpy.diag((w/h)*c(xd))
17          b_loc=numpy.diag(w*b(xd))
18          a_loc=numpy.diag((h*w)*a(xd))
19          Al=(DT @ c_loc+ET @ b_loc) @ D+ET @ a_loc @ E
20          row=l*m+i_loc_sp_mat
21          col=l*m+j_loc_sp_mat
22          A=A+sparse.coo_matrix((Al.flatten(),(row, col)), \
23              shape=(n_total,n_total))
24          Fl=ET @ (h*w*f(xd)).reshape(—1,1)
25          F=F+sparse.coo_matrix((Fl.flatten(), \
26              (l*m+i_loc, numpy.zeros(m+1))),shape=(n_total,1))
27      return A,F
```

Listing 2.17: Python code for assembling one-dimensional matrices.

Script for assembling BC (see listing 2.18) follows linear logic of replacing corresponding values in stiffness matrices by the BC values. Some comments are added to make it clear which part of the code responsible for the BC of the first type and which one for the third type. The function returns matrices A0 and F0 which are ready to be used with linear solver.

```
1 def assemblingBC(bc,A,F):
2       # bc - boundary conditions matrix of size 2x3 :
3       # bc(1,:)=[1,0,u_s0] , or bc(1,:)=[3,sig_s0,mu_s0]
4       # bc(2,:)=[1,0,u_s1] , or bc(2,:)=[3,sig_s1,mu_s1]
5       n=F.shape[0]—1 #number of all mesh points
6       # set boundary conditions of 3-d type
```

```
7    if bc[0,0]==3:
8        A[0,0]=A[0,0]+bc[0,1]
9        F[0]=F[0]+bc[0,2]
10   if bc[1,0]==3:
11       A[n,n]=A[n,n]+bc[1,1]
12       F[n,0]=F[n,0]+bc[1,2]
13   #set boundary conditions of the first type
14   ib=0; ie=n+3
15   if bc[0,0]==1:
16       ib=1; u_s0=bc[0,2]
17       F=F—u_s0*A[:,0]
18   if bc[1,0]==1:
19       ie=n+2; u_s1=bc[1,2]
20       F=F—u_s1*A[:,n]
21   A0=A[ib:ie,ib:ie]
22   F0=F[ib:ie,0]
23   return A0,F0
```

Listing 2.18: Python code for assembling one-dimensional boundary conditions.

Listings 2.19 and 2.20 show function implementations for solving FEM and BVP. Note how the sparse linear system is solved using **sparse**.linalg.spsolve() function.

```
1 def solveFEM(bc,A0,F0):
2     u_s0 =[]; u_s1 =[]
3     if bc[0,0]==1: u_s0=bc[0,2]
4     if bc[1,0]==1: u_s1=bc[1,2]
5     u0=sparse.linalg.spsolve(A0,F0)
6     u=u_s0
7     u=numpy.append(u,u0)
8     u=numpy.append(u,u_s1)
9     return u
```

Listing 2.19: Python code for solving one-dimensional FEM system.

```
1 def solveBVP(bc,c,b,a,f,xl,x,d,w):
2     m=numpy.size(x)—1
3     n=numpy.size(xl)
4     N=(n—1)*m+1 # number of mesh points
5     x_mesh=numpy.zeros((N,)) # all mesh points
```

```
6       for l in range(n—1):
7           ne=l*m+numpy.arange(m+1) # mesh point indices
8           x_mesh[ne]=xl[l]+(xl[l+1]—xl[l])*x
9       A,F=assemblingAF(c,b,a,f,xl,x,d,w)
10      A0,F0=assemblingBC(bc,A,F)
11      u=solveFEM(bc,A0,F0)
12      return u,x_mesh
```

Listing 2.20: Python code for solving one-dimensional BVP.

Finally, the main script for testing example problem is given in listing 2.21. To specify equation coefficients and exact solution, we use in-line function definitions (lines 6–12) with corresponding vectorized numpy functions. Also, built-in function numpy.linalg.norm() is used to calculate various norms, for example, $\|\cdot\|_\infty$ norm is printed in lines 37–39.

```
1   def mainTest():
2       # set BVP data:
3       # integration domain
4       s_0=—1; s_1=1
5       # exact solution
6       def u(x): return numpy.sin(x)
7       def du(x): return numpy.cos(x)
8       # equation coefficients
9       def c(x): return numpy.cos(x)
10      def b(x): return —2*numpy.sin(x)
11      def a(x): return numpy.sin(x)
12      def f(x): return numpy.sin(x)**2
13      # boundary conditions
14      sigma=0
15      mu=c(s_1)*du(s_1)+sigma*u(s_1)
16      bc=numpy.array([[1, 0, u(s_0)],
17                      [3, sigma, mu] ])
18      # set FEM parameters
19      m=3 # polynomial degree
20      ne=6 # number of final elements
21      M=m+1 # number of quadrature nodes
22      xl=numpy.linspace(s_0,s_1,ne+1) # finite element nodes
23      t,_ =LobattoQuad(m+1,0,1) # interpolation nodes
24      d,w=LobattoQuad(M,0,1) # quadrature nodes and weighs
25      # solve BVP
26      start_time = timeit.default_timer()
```

```
27    uh,x=solveBVP(bc,c,b,a,f,xl,t,d,w)
28    print("solving time =",timeit.default_timer() - start_time)
29    # find error in L_2 and H^1 norms
30    e0,e1=errorNorm(u,du,uh,xl,t)
31    print("L2 norm error =",e0)
32    print("H1 norm error =",e1)
33    X,Uh,dUh=getRefinedValues(uh,xl,t,10*m)
34    # find error in various norms
35    Z=u(X)-Uh; z=u(x)-uh; dZ=du(X)-dUh
36    ie=numpy.arange(0,numpy.size(x),m)
37    print("uh error inf norm:",numpy.linalg.norm(Z,numpy.inf))
38    print("duh error inf norm:",numpy.linalg.norm(dZ,numpy.inf))
39    print("error at f.e. nodes inf norm:", \
40        numpy.linalg.norm(u(xl)-uh[ie],numpy.inf))
41    plt.figure()
42    # plot solution
43    plt.plot(X,Uh,'-b',x,uh,'rx',x[ie],uh[ie],'ro')
44    plt.xlabel('x'); plt.ylabel('u_h')
45    plt.show()
46    #plot solution error
47    plt.plot(X,Z,'-b',x,z,'rx',x[ie],z[ie],'ro')
48    plt.xlabel('x'); plt.ylabel('u-u_h')
49    plt.show()
50    #plot derivatives error
51    plt.plot(X,dZ,'-b', \
52        x,numpy.zeros(numpy.size(x)),'rx', \
53        x[ie],numpy.zeros(numpy.size(x[ie])),'ro')
54    plt.xlabel('x'); plt.ylabel('u''-u''_h');
55    plt.show()
```

Listing 2.21: Python code for solving example one-dimensional BVP.

CHAPTER 3

Finite Element Method for the Two-Dimensional Boundary Value Problem

B Y learning how to solve one-dimensional BVP using FEM, we have gained knowledge allowing us to take next step and consider two-dimensional case. Although all steps of the algorithm for the two-dimensional case are similar to the one-dimensional one, their implementations are different and require specific features and coding strategies to be considered.

We begin by formulating a model problem in terms of linear partial differential equation with mixed boundary conditions. Then, an integral formulation of the model problem is given with further finite element discretization. We introduce basis functions and show how the discretized problem can be reduced to solving a linear system. Also, basic concepts of two-dimensional mesh with finite elements having triangular shape are discussed.

Notations introduced in chapter 1 are used across the current chapter along with new ones given below.

DOI: 10.1201/9781003265979-3

GLOSSARY

$f(x)$: a function of multiple variables.

$\boldsymbol{f}(\boldsymbol{x})$: a multivariate function of multiple variables.

$\frac{\partial f}{\partial x_1}, \frac{\partial f}{\partial x_2}, \frac{\partial^2 f}{\partial x_1 \partial x_2}, \dots$: partial derivatives of a function $f(x)$.

∇: the nabla operator with definition $\nabla f = \left(\frac{\partial f}{\partial x_1}, \frac{\partial f}{\partial x_2} \right)^T$.

3.1 MODEL PROBLEM

Similar to the one-dimensional case, two-dimensional FEM approach will be based on considering a generic BVP which can be reduced to a specific problem by simplifying particular terms in the BVP formulation. This allows to deduce a FEM algorithm which covers as many application problems as possible.

Consider a BVP of finding a function $u = u(\boldsymbol{x})$ in two-dimensional region Ω with piecewise smooth boundary $\Gamma = \Gamma_D \cup \Gamma_R$

$$-\boldsymbol{\nabla} \cdot (c(\boldsymbol{x})\nabla u) + \boldsymbol{b}(\boldsymbol{x}) \cdot \nabla u + a(\boldsymbol{x})u = f(\boldsymbol{x}), \quad \boldsymbol{x} = (x_1, x_2) \in \Omega, \quad (3.1)$$

$$u(\boldsymbol{x}) = u_D(\boldsymbol{x}), \; \boldsymbol{x} \in \Gamma_D,$$
$$c(\boldsymbol{x})\nabla u \cdot \boldsymbol{\nu}(\boldsymbol{x}) + \sigma(\boldsymbol{x})u = \mu(\boldsymbol{x}), \; \boldsymbol{x} \in \Gamma_R.$$

Here, Γ_D and Γ_R are the boundary parts for Dirichlet and Robin boundary conditions, respectively, and either of them could be empty (in this case the corresponding boundary condition is omitted); Γ is supposed to be closed; $\boldsymbol{\nu}(\boldsymbol{x})$ is the unit outward normal vector at the point $\boldsymbol{x}$. Functions c, a, f, u_D, σ, μ and $\boldsymbol{b}(\boldsymbol{x}) = (b_1(\boldsymbol{x}), b_2(\boldsymbol{x}))^T$ are known and supposed to be piecewise smooth. The model problem (3.1) can be reduced to variety of particular problems with appropriate definition of the coefficients, for example, when $c(\boldsymbol{x}) = 1$, $\boldsymbol{b}(\boldsymbol{x}) = a(\boldsymbol{x}) = 0$ we come to a Poisson's equation.

Let us reformulate the problem (3.1) in terms of integral equations. First, we consider function sets $\mathcal{W}$ and $\mathcal{W}^0$

$$\mathcal{W} = \left\{ w \in H^1(\Omega) : w(\boldsymbol{x}) = u_D(\boldsymbol{x}), \; x \in \Gamma_D \right\},$$
$$\mathcal{W}^0 = \left\{ w \in H^1(\Omega) : w(\boldsymbol{x}) = 0, \; x \in \Gamma_D \right\},$$

where $H^1(\Omega)$ is the Sobolev space. Then, perform a standard sequence of steps to come to the integral formulation of the problem. We follow the logic of the projective method described in section 1.2 which is equally applied to the two-dimensional case. By multiplying equation (3.1) by function $v \in \mathcal{W}^0$, integrating over region Ω, applying Gauss theorem[1] and taking into account boundary conditions we come to the formulation of the problem in the form

$$\int_\Omega (c\nabla u \cdot \nabla v + \boldsymbol{b} \cdot \nabla u v + auv)\, d\boldsymbol{x} + \int_{\Gamma_R} \sigma uv d\boldsymbol{x}$$

$$= \int_\Omega fv d\boldsymbol{x} + \int_{\Gamma_R} \mu v d\boldsymbol{x}. \quad (3.2)$$

Now, the problem can be formulated as follows: find a function $u \in \mathcal{W}$ so that, for any $v \in \mathcal{W}^0$, the function u satisfies equation (3.2). Here, functions v are the test functions.

3.2 FINITE ELEMENTS DEFINITION

Let us introduce $\mathcal{T}_h$ - partitioning (triangulation) of the domain Ω by finite elements τ, and all boundary points Γ are vertices of some elements. A finite element is a triple $(\tau, \mathcal{P}_\tau, \mathcal{G}_\tau)$, where τ is the closed area of a simple form (triangle or polygon), $\mathcal{P}_\tau$ are the interpolation points belonging to τ and $\mathcal{G}_\tau$ are interpolation functions defined on τ (maximum diameter of elements in $\mathcal{T}_h$ is denoted by h). Then, Γ_D^h and Γ_R^h are partitioning of Γ_D and Γ_R, respectively. We introduce notation: $\mathcal{P}_h$ is a set of all interpolation points (grid points of $\mathcal{T}_h$), $\mathcal{P}_D$ is a set of points belonging to Γ_D^h and $\mathcal{P}_R$ is a set of points belonging to Γ_R^h. Now, we are looking for functions u_h and v_h defined on grid $\mathcal{P}_h$ so that

$$\int_{\mathcal{T}_h} (c\nabla u_h \cdot \nabla v_h + \boldsymbol{b} \cdot \nabla u_h v_h + au_h v_h)\, d\boldsymbol{x} + \int_{\Gamma_R^h} \sigma u_h v_h d\boldsymbol{x}$$

$$= \int_{\mathcal{T}_h} fv_h d\boldsymbol{x} + \int_{\Gamma_R^h} \mu v_h d\boldsymbol{x}. \quad (3.3)$$

Function u_h represents an approximate solution to the BVP and the FEM is to find this function.

---

[1] Gauss theorem: $\int_\Omega \nabla \cdot \boldsymbol{b} v d\boldsymbol{x} = -\int_\Omega \boldsymbol{b} \cdot \nabla v d\boldsymbol{x} + \int_\Gamma \boldsymbol{b} \cdot \boldsymbol{\nu} v d\boldsymbol{x}.$

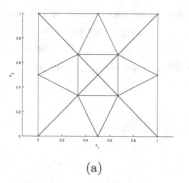
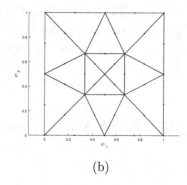

(a)   (b)

FIGURE 3.1: Examples of the triangulation with three (a) and six (b) interpolation points over a triangle element.

3.3   TRIANGULATION EXAMPLES

Figure 3.1(a) shows example of the domain triangulation with three interpolation points. In this case, the finite element τ is a triangle, interpolation points $\mathcal{P}_\tau$ are the triangle's vertices and interpolation functions $\mathcal{G}_\tau = \mathcal{G}_1 = \{g_1 + g_2 x_1 + g_3 x_2\}$ are polynomials of the first order. Any two elements share either edge or vertex and each edge contains two interpolation points.

Figure 3.1(b) shows triangulation with six interpolation points. Each $\mathcal{G}_\tau$ includes vertices and in-between points belonging to the edges. As a result, three points of interpolation are available for each edge and the interpolation function is of the second order, $\mathcal{G}_\tau = \mathcal{G}_2 = \{g_1 + g_3 x_2 + g_4 x_1^2 + g_5 x_1 x_2 + g_6 x_2^2\}$.

3.4   LINEAR SYSTEM OF THE FEM

Equation (3.3) reduces to the system of linear equation. Firstly, we need to select basis functions which define interpolation over finite elements τ. We define basis functions as follows: Let us enumerate all points $\mathcal{P}_h$ from 1 to n_p and assign a basis function $\varphi_i(\boldsymbol{x})$ to each point $\boldsymbol{p}_i \in \mathcal{P}_h$ so that $\varphi_i(\boldsymbol{p}_j) = \delta_{ij}$, $j = 1, \ldots, n_p$. [2] Then, any v_h can be represented in

---

[2] $\delta_{ij} = 1$ if $i = j$ and $\delta_{ij} = 0$ if $i \neq j$. Basis function φ has an important property: if $\boldsymbol{p}_i$ does not belong to a finite element then $\varphi(\boldsymbol{p}_i) = 0$ on this element.

the form of interpolating polynomial

$$v_h(x) = \sum_{i=1}^{n_p} v_i \varphi_i(x),$$

where $v_i = v(p_i)$. Let us divide set of indices $\mathcal{I} = \{1, 2, \ldots, n_p\}$ into two subsets: $\mathcal{I}_D = \{i \in \mathcal{I} : p_i \in \mathcal{P}_D\}$ and $\mathcal{I}_U = \{i \in I : p_i \notin \mathcal{P}_D\}$. Solution at the points p_{i_D} is known from the boundary condition $(u_h(p_{i_D}) = u_{D_i})$. We need to find a solution at the points p_{i_U}.

The following expansions are valid in $\mathcal{T}_h$:

$$u_h(x) = \sum_{i \in i_U} u_i \varphi_i(x) + \sum_{i \in i_D} u_{D_i} \varphi_i(x), \quad v_h(x) = \sum_{i \in i_U} v_i \varphi_i(x). \quad (3.4)$$

Let us define a matrix $\tilde{A} \sim (n_p \times n_p)$ and a vector $\tilde{F} \sim (n_p \times 1)$ having components

$$\tilde{A}_{ij} = \int_{\mathcal{T}_h} (c\nabla\varphi_j \cdot \nabla\varphi_i + b \cdot \nabla\varphi_j \varphi_i + a\varphi_j\varphi_i) \, dx + \int_{\Gamma_R^h} \sigma\varphi_j\varphi_i dx \quad (3.5)$$

$$\tilde{F}_i = \int_{\mathcal{T}_h} f(x)\varphi_i dx + \int_{\Gamma_R^h} \mu(x)\varphi_i dx \quad (3.6)$$

where $\tilde{A}$ is called global stiffness matrix and $\tilde{F}$ is the global forcing vector. Then, equation (3.3) can be rewritten in the form of the linear system (see section 1.3.3 for details)[3]

$$\sum_{j \in \mathcal{I}_U} \tilde{A}_{ij} u_j = q_i, \quad q_i = \tilde{F}_i - \sum_{j \in \mathcal{I}_D} \tilde{A}_{ij} u_{D_j}, \quad i \in \mathcal{I}_U. \quad (3.7)$$

By solving system (3.7) we define u_j, $j \in \mathcal{I}_U$, and the approximation solution of the problem (3.2) is found from (3.4). Although matrix and vector elements can be defined from (3.5) and (3.6), these formulas are not usually used for calculation purpose. There exists more convenient and efficient calculation method which we consider below.

---

[3]Equation (3.3) follows from equation (3.7). This can be seen if we multiply i^{th} equation in (3.7) by v_i and sum up all equations taking into account (3.4)–(3.6).

3.5 STIFFNESS MATRIX AND FORCING VECTOR

Linear system (3.7) implies calculation of matrix $\tilde{A}$ and vector $\tilde{\boldsymbol{F}}$. In practical application problems, these matrices are usually of large sizes and chosen calculation method affects overall performance of the FEM programming implementation. Here, we consider efficient method of their calculation.

It is followed from (3.5) that matrix $\tilde{A}$ is a sum of two square matrices of $(n_p \times n_p)$ dimension: $\tilde{A} = A + S$,[4]

$$A_{ij} = \int_{\mathcal{T}_h} (c\nabla\varphi_j \cdot \nabla\varphi_i + \boldsymbol{b} \cdot \nabla\varphi_j\varphi_i + a\varphi_j\varphi_i)\, d\boldsymbol{x},$$

$$S_{ij} = \int_{\Gamma_R^h} \sigma\varphi_j\varphi_i d\boldsymbol{x}, \quad i,j = 1, \ldots, n_p.$$

Similarly,

$$\tilde{\boldsymbol{F}} = \boldsymbol{F} + \boldsymbol{M}, \quad F_i = \int_{\mathcal{T}_h} f\varphi_i d\boldsymbol{x}, \quad M_i = \int_{\Gamma_R^h} \mu\varphi_i d\boldsymbol{x}, \quad i = 1, \ldots, n_p.$$

Recall that triangulation $\mathcal{T}_h$ is a union of finite elements τ, that is $\mathcal{T}_h = \bigcup\limits_{l=1}^{n_t} \tau_l$ where n_t is the total number of finite elements. It is assumed that there are m_τ number of interpolation points belonging to the element τ_l (for the triangular $\mathcal{G}_1$ element $m_\tau = 3$) and these points are ordered (locally numerated): $\boldsymbol{p}_{i_1}, \boldsymbol{p}_{i_2}, \ldots, \boldsymbol{p}_{i_{m_\tau}}$. Let us specify a connectivity matrix, $t \sim (m_\tau \times n_t)$, and place into l^{th} column the indices of the points of the lth element $(i_1, i_2, \ldots, i_{m_\tau})$ so that t_{jl} defines index of the jth point (in local order) belonging to the element τ_l.

Calculation of A and F. Represent the integral over triangulation $\mathcal{T}_h$ through the sum of integrals over elements τ_l. Using the matrix of indices, t_{jl}, we can introduce a matrix $\bar{A}^l$ with the components

$$\bar{A}^l_{\alpha\beta} = \int_{\tau_l} (c\nabla\varphi_{t_{\beta l}} \cdot \nabla\varphi_{t_{\alpha l}} + \boldsymbol{b} \cdot \nabla\varphi_{t_{\beta l}}\varphi_{t_{\alpha l}} + a\varphi_{t_{\beta l}}\varphi_{t_{\alpha l}})\, d\boldsymbol{x},$$

$$\alpha, \beta = 1, \ldots, m_\tau,$$

---

[4]Note that S is a symmetric matrix; A is a symmetric matrix provided $\boldsymbol{b} = 0$.

which is the local stiffness matrix of the finite element τ_l. Now, we can express global matrix in terms of local stiffness matrices by introducing extended local stiffness matrix $A^l \sim (n_p \times n_p)$ so that matrix A^l contains zeros except of elements with indices $t_{\alpha l}$ and $t_{\beta l}$, and

$$A^l_{t_{\alpha l}, t_{\beta l}} = \bar{A}^l_{\alpha\beta}, \quad \alpha, \beta = 1, \ldots, m_\tau.$$

For example, matrix A^l for the triangular $\mathcal{G}_1$ element has $m_\tau \times m_\tau = 3 \times 3 = 9$ nonzero elements. As a result,

$$A = \sum_{l=1}^{n_t} A^l.$$

The latter expression indicates that the global stiffness matrix A can be obtained by summation of the extended local stiffness matrices, A^l. Since there is no sense to sum up zeros, we come to the algorithm of composing global matrix A (see algorithm 4).

Algorithm 4 Composing global stiffness matrix A

---

1. Specify connectivity matrix of indices $t \sim (m_\tau \times n_t)$.

2. Define zero value matrix $A \sim (n_p \times n_p)$.

3. For each $l = 1, 2, \ldots, n_t$:

   (a) calculate matrix $\bar{A}^l_{\alpha\beta} \sim (m_\tau \times m_\tau)$

   (b) perform summation across indices $t_{\alpha l}$ and $t_{\beta l}$: $A_{t_{\alpha l}, t_{\beta l}} = A_{t_{\alpha l}, t_{\beta l}} + \bar{A}^l_{\alpha\beta}, \ \alpha, \beta = 1, 2, \ldots, m_\tau$

Similarly, the algorithm for calculation of F can be obtained (see algorithm 5). Vector $\bar{F}^l$ is called the local force vector.

Calculation S and M. Matrix S and vector M contain boundary integrals over Γ_R^h, so the calculations are performed only using interpolation points belonging to the boundary. Let us introduce, analogous to matrix of indices t, a matrix of indices associated with edges, $e \sim (m_e \times n_e)$, which is called edge matrix. Here n_e is the number of edges belonging to the boundary Γ_R^h and m_e is the number of interpolation points on the edge. In other words, e_{jk} defines index of the jth point (in local order) belonging to the kth boundary edge. As a result,

---

Algorithm 5 Composing global vector $\boldsymbol{F}$

---

1. Define zero value vector $\boldsymbol{F} \sim (n_p \times 1)$.

2. For each $l = 1, 2, \ldots, n_t$:

   (a) calculate $\bar{\boldsymbol{F}}^l \sim (m_\tau \times 1)$:

   $$\bar{F}^l_\alpha = \int_{\tau_l} f(\boldsymbol{x}) \varphi_{t_{\alpha l}}(\boldsymbol{x}) d\boldsymbol{x}$$

   (b) perform summation across indices $t_{\alpha l}$: $F_{t_{\alpha l}} = F_{t_{\alpha l}} + \bar{F}^l_\alpha$, $\alpha = 1, 2, \ldots, m_\tau$

---

for the kth edge, e_k, we can define a local matrix $\bar{S}^k$ with elements

$$\bar{S}^k_{\alpha\beta} = \int_{e_k} \sigma(\boldsymbol{x}) \varphi_{e_{\beta k}} \varphi_{e_{\alpha k}} d\boldsymbol{x}.$$

By introducing extended matrix S^k (see similar extended matrix A^l) we obtain

$$S = \sum_{k=1}^{n_e} S^k.$$

Vector $\boldsymbol{M}$ is calculated through introducing a local vector $\bar{\boldsymbol{M}}^k$ for each edge e_k

$$\bar{M}^k_\alpha = \int_{e_k} \mu(\boldsymbol{x}) \varphi_{e_{\alpha k}}(\boldsymbol{x}) d\boldsymbol{x},$$

and summation of the extended vectors $\boldsymbol{M}^k$ across all edges

$$\boldsymbol{M} = \sum_{k=1}^{n_e} \boldsymbol{M}^k.$$

Algorithm of composing matrix S and vector $\boldsymbol{M}$ is the same as for matrix A and vector $\boldsymbol{F}$.

3.6 ALGORITHM OF SOLVING PROBLEM

Algorithm 6 summarizes previous results and describes sequence of steps to find the approximate solution to the problem (3.2).

Algorithm 6 Generic two-dimensional FEM

---

1. Define geometry of the problem (domain Ω and their boundaries Γ_D and Γ_R);

2. Define boundary conditions;

3. Build triangulation of the domain Ω;

4. Define a connectivity matrix, t, and an edge indices matrix, e;

5. Compose matrix A and vector $\boldsymbol{F}$;

6. Define indices i_U and i_D;

7. Compose matrix S and vector $\boldsymbol{M}$;

8. Compose matrix $\tilde{A}$ and vector $\boldsymbol{q}$ (see (3.7));

9. Solve linear system (3.7) and find the vector of nodal values $\boldsymbol{u}$ for the solution u_h ;

10. Express solution u_h through basis functions expansion;

In the following chapters we will discuss each step of this algorithm and provide implementation codes which form a FEM framework library. Programming framework for the FEM will be obtained in MATLAB as well as in Python programming language. Mathematical formulations obtained above will serve as defining expressions for coding numerical methods. To make the code consistent, the same data structures for the FEM representation across both languages implementations are going to be used.

CHAPTER 4

Building Two-Dimensional Meshes

F IRST four steps of the FEM algorithm given in section 3.6 will be covered in this chapter. We will describe how to define geometry of a problem and build two-dimensional triangular mesh using MATLAB and Python tools. Basic data structure for the mesh representation will be considered.

Basically, generating geometry for a domain with simple boundary shape is not a challenge; however things become more complicated when a geometry has subdomains or nontrivial boundary shape. In general, the problem of generating geometry can be viewed as having two tasks: geometry definition through some data structure, and performing triangulation to build finite elements. In this chapter we are focussed on the first task, namely, various approaches of how to create data structures for geometry definition and how to represent mesh in a matrix form. Questions of how to encode boundary and build geometry will be discussed as well.

To perform triangulation itself using predefined geometry definition, we are going to use specific MATLAB and Python tools. Triangulation algorithm implementation is not discussed below and it is beyond the scope of the book, but rather existing implementation as a tool or third-party library is used. Interested reader can find mesh generating algorithms in [3] where some triangulation techniques are discussed.

DOI: 10.1201/9781003265979-4

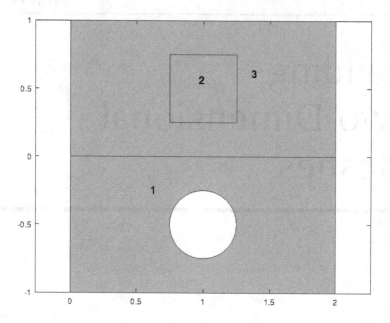

FIGURE 4.1: Example of the domain with labelled subdomains.

4.1 DEFINING GEOMETRY

Let us start with considering an example geometry to show how it can be represented to use with the FEM solver. Figure 4.1 shows an example of the geometry domain Ω. This domain consists of three labelled subdomains, Ω_1, Ω_2 and Ω_3 so that $\Omega = \Omega_1 \cup \Omega_2 \cup \Omega_3$. It is seen that Ω_1 and Ω_3 are rectangles and Ω_2 is a square. A circle of radius 0.25 is removed from Ω_1. Also, it is highlighted that we have internal area of the geometry (coloured in grey) and external one which is called complement of the domain Ω and will be referenced as Ω_0. The geometry is split into subdomains to illustrate the fact that PDE coefficients could be different in these subdomains and we need to take it into account.

Next, boundaries of the geometry are considered (see figure 4.2). Boundaries of Ω_1, Ω_2 and Ω_3 are split into segments. Segments are labelled, they cannot intersect and have single common end point. As a result, the square is specified with a minimum of four segments and the circle – with two segments. We define internal and external (boundary) segments. Specific boundary conditions could be specified on each seg-

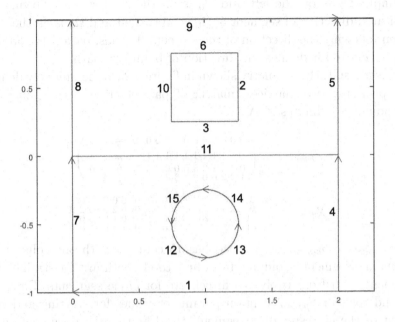

FIGURE 4.2: Example domain with labelled boundaries.

ment. Figure 4.2 shows the boundary consisting of 15 segments including 5 internal ones (labelled as 2, 3, 6, 10 and 11). Each segment line can be described in a parametric form

$$x_1 = x_1(s), \quad x_2 = x_2(s), \quad s \in [s_0, s_1].$$

For example, a straight line can be parameterized as follows:

$$\boldsymbol{x}(s) = \boldsymbol{x}^{start} + s\left(\boldsymbol{x}^{end} - \boldsymbol{x}^{start}\right), \quad s \in [0, 1], \tag{4.1}$$

where $\boldsymbol{x}^{start} = \boldsymbol{x}(0)$ and $\boldsymbol{x}^{end} = \boldsymbol{x}(1)$ are starting and ending points of a line segment. A circle segment has a parametric form

$$x_1(s) = r\cos(s) + x_1^{centre}, \quad x_2(s) = r\sin(s) + x_2^{centre}, s \in [s_0, s_1], \tag{4.2}$$

where r is a radius and $\boldsymbol{x}^{centre}$ is a circle centre point.

When following the segment from the starting point having parameter value of s_0 to the endpoint with parameter value of s_1, there is always one subdomain on the left and one on the right hand side. For example, Ω_3 is on the left and Ω_0 is on the right when following the indicated direction of segment 5; Ω_0 is on the left and Ω_3 is on the right when following the direction of the segment 9. Thus, parameter order is important and it defines the direction of boundary path.

As a result, the geometry shown in figure 4.2 can be uniquely defined by specifying two matrices: matrix of parameter values, S_{param}, and endpoints coordinates $X_1 X_2$

$$S_{param} = \begin{pmatrix} 0 & 0 & 0 & 0 & 0 & 0 & 0 & 0 & 0 & 0 & 0 & \pi & \frac{3\pi}{2} & 0 & \frac{\pi}{2} \\ 1 & 1 & 1 & 1 & 1 & 1 & 1 & 1 & 1 & 1 & 1 & \frac{3\pi}{2} & 2\pi & \frac{\pi}{2} & \pi \\ 0 & 3 & 3 & 1 & 3 & 3 & 0 & 0 & 0 & 3 & 3 & 0 & 0 & 0 & 0 \\ 1 & 2 & 2 & 0 & 0 & 2 & 1 & 3 & 3 & 2 & 1 & 1 & 1 & 1 & 1 \end{pmatrix}$$

$$X_1 X_2 = \begin{pmatrix} 2 & 1.25 & 1.25 & 2 & 2 & 0.75 & 0 & 0 & 0 & 0.75 & 0 \\ 0 & 1.25 & 0.75 & 2 & 2 & 1.25 & 0 & 0 & 2 & 0.75 & 2 \\ -1 & 0.75 & 0.25 & -1 & 0 & 0.75 & -1 & 0 & 1 & 0.25 & 0 \\ -1 & 0.25 & 0.25 & 0 & 1 & 0.75 & 0 & 1 & 1 & 0.75 & 0 \end{pmatrix}$$

Here, matrix $S_{param} \sim (4 \times 15)$ has 15 columns with one column per boundary segment. Columns from first to eleventh are for straight line segments and from twelve to fifteen are for circle segments. First and second rows of S_{param} contain parameter values for starting and endpoint of the corresponding segment, third and fourth rows contain labels of the left hand and right hand regions correspondingly. Matrix $X_1 X_2 \sim (4 \times 11)$ encodes coordinates of straight line segments with one column per segment. Here, first two rows are x_1 coordinates of starting and endpoints, and the last two rows are x_2 coordinates of line segments correspondingly. Having matrices S_{param} and $X_1 X_2$ along with formulas (4.1) and (4.2) allows calculating coordinates of any point on the boundary given an index of boundary segment. In general, quite complicated geometries (not only lines and circles) can be represented through parametric form and encoded in the similar way.

Here, we introduce a MATLAB function geometryFunction(b_index , s) which returns (x_1, x_2) coordinates of a boundary point for any segment with index b_index and parameter value s. Parameter b_index can be either scalar or vector and it is allowed to omit parameters. Function call with a single argument, geometryFunction(b_index), returns corresponding entries of the matrix S_{param} with one column for each boundary segment specified in b_index. If the function is called without arguments, geometryFunction(), it returns total number of boundary segments.

```
1  function [x1,x2]= geometryFunction (b_index, s)
2    % Geometry data
3    s_param=[ 0 0 0 0 0 0 0 0 0 0 0     pi 3/2*pi    0 pi/2
4              1 1 1 1 1 1 1 1 1 1 1 3/2*pi   2*pi pi/2   pi
```

```
 5          0 3 3 1 3 3 0 0 0 3 3     0     0    0    0
 6          1 2 2 0 0 2 1 3 3 2 1     1     1    1    1 ];
 7  x1x2=[2 1.25 1.25  2 2 0.75  0 0 0 0.75 0
 8        0 1.25 0.75  2 2 1.25  0 0 2 0.75 2
 9       -1 0.75 0.25 -1 0 0.75 -1 0 1 0.25 0
10       -1 0.25 0.25  0 1 0.75  0 1 1 0.75 0 ];
11  if nargin==0, x1=size(s_param,2); return; end
12  if nargin==1, x1=s_param(:,b_index); return; end
13  x1=zeros(size(s));
14  x2=zeros(size(s));
15  [m,n]=size(b_index);
16  if m==1 && n==1, b_index=b_index*ones(size(s)); end
17  if ~isempty(s),
18    % line segments
19    for k=1:11
20      i=find(b_index==k);
21      if ~isempty(i)
22        x1(i)=x1x2(1,k)+(x1x2(2,k)-x1x2(1,k))*s(i);
23        x2(i)=x1x2(3,k)+(x1x2(4,k)-x1x2(3,k))*s(i);
24      end
25    end
26    % circle segments
27    for k=12:15
28      i=find(b_index==k);
29      if ~isempty(i)
30        c_x1=1; c_x2=-0.5; rc=0.25;
31        x1(i)=rc*cos(s(i))+c_x1;
32        x2(i)=rc*sin(s(i))+c_x2;
33      end
34    end
35  end
36 end
```

Listing 4.1: MATLAB code for generating geometry boundaries.

After defining geometry boundaries, the next step is to perform triangulation to obtain a mesh for the FEM algorithm. Before looking into specific methods of mesh building, let us first discuss data structures which are used for triangular mesh representation.

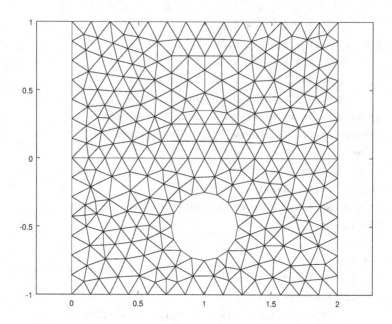

FIGURE 4.3: Mesh for example geometry.

4.2 REPRESENTING MESHES IN MATRIX FORM FOR LINEAR IN-TERPOLATION FUNCTIONS

Above, we have discussed data structures to encode geometry with sub-domaines definition given in figure 4.1 and boundary segments specifi-cation given in figure 4.2. The next logical step would be considering triangulation which is illustrated in figure 4.3 for the example geometry. In this book, only triangular mesh is used leaving other types beyond the scope (for example, mesh with rectangular elements).

In this section, we do not consider methods to build a mesh itself but instead, we are focussed on how the mesh can be represented as a data structure. It is assumed that there exists a triangular mesh and we want to encode it. Linear interpolation functions are considered here (with three interpolation nodes), so triangle vertices are the mesh nodes and interpolation nodes at the same time. Let us call it $\mathcal{G}_1$ mesh stressing that it is used when solution is approximated by a polynomial of the first order.

In general, the mesh data structure should serve two main purposes. Firstly, it should provide information about triangle vertices to allow performing interpolation on elements. Secondly, it should contain information about element edges which belong to boundary segments to correctly apply BC. As such, the mesh data should encapsulate information about mesh points, their connectivity to form triangles (finite elements) and relation between element edges and boundary segments. One of the most convenient forms of mesh representation which has became an industry standard is a matrix form. Three matrices, P, T and E, store all necessary information which allow apply FEM interpolation algorithms. Below we discuss them individually.

Given enumerated n_p mesh points, let us form a matrix $P \sim (2 \times n_p)$ which stores coordinates of mesh points (triangle vertices) so that for i^{th} point coordinate $x_1 = P_{1,i}$ and $x_2 = P_{2,i}$. Matrix P can be used to access any mesh points but it does not keep any information about mesh elements formation.

Structure representing how mesh triangles are formed from the set of vertices is called connectivity matrix, T. Given a number of triangles, n_t, matrix $T \sim (4 \times n_t)$ stores indices of vertices for each triangular element and corresponding subdomain label to which this triangle belongs. Elements $T_{1,j}, T_{2,j}$ and $T_{3,j}$ are the indices of vertices of the jth triangle (counted anticlockwise), element $T_{4,j}$ is the subdomain label. Note that the matrix T is formed using indices of points from matrix P, so the order in which points are arranged in P is important.

The last structure is an edge matrix, E, which stores information regarding finite element edges belonging to the boundary segments (internal and external). Given a number of edges, n_e, matrix $E \sim (7 \times n_e)$ contains the following information. Elements $E_{1,k}$ and $E_{2,k}$ contain indices of vertices belonging to the kth edge (vertex index with respect to matrix P), elements $E_{3,k}$ and $E_{4,k}$ are the corresponding parameter values s_0 and s_1 for parametric representation, element $E_{5,k}$ is the index of boundary segment to which the edge belongs, elements $E_{6,k}$ and $E_{7,k}$ are the subdomain labels from the left and right hand side of the edge according to the parameterization direction.

To better understand mesh encoding, let us consider a simple case of L-shaped geometry given in figure 4.4 with 3 subdomains and 10 boundary segments (8 external and 2 internal segments). Splitting each square (subdomain) into two triangles results in rough mesh with 6 finite elements. In figure 4.4, all points and boundary segments are enumerated.

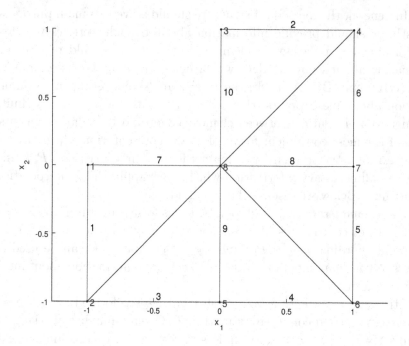

FIGURE 4.4: L-shaped domain with rough triangulation.

Matrices P, T and E for the mesh shown in figure 4.4 have the form

$$P = \begin{pmatrix} -1 & -1 & 0 & 1 & 0 & 1 & 1 & 0 \\ 0 & -1 & 1 & 1 & -1 & -1 & 0 & 0 \end{pmatrix},$$

$$T = \begin{pmatrix} 1 & 4 & 7 & 2 & 5 & 6 \\ 2 & 3 & 4 & 5 & 6 & 7 \\ 8 & 8 & 8 & 8 & 8 & 8 \\ 2 & 1 & 1 & 2 & 3 & 3 \end{pmatrix},$$

$$E = \begin{pmatrix} 1 & 3 & 2 & 5 & 6 & 7 & 1 & 8 & 5 & 8 \\ 2 & 4 & 5 & 6 & 7 & 4 & 8 & 7 & 8 & 3 \\ 0 & 0 & 0 & 0 & 0 & 0 & 0 & 0 & 0 & 0 \\ 1 & 1 & 1 & 1 & 1 & 1 & 1 & 1 & 1 & 1 \\ 1 & 2 & 3 & 4 & 5 & 6 & 7 & 8 & 9 & 10 \\ 2 & 0 & 2 & 3 & 3 & 1 & 0 & 1 & 2 & 0 \\ 0 & 1 & 0 & 0 & 0 & 0 & 2 & 3 & 3 & 1 \end{pmatrix}.$$

A reader is encouraged to examine matching entries of these matrices to the mesh element labels in figure 4.4.

4.3   COMPLEMENTARY MESH

Representation of the mesh through matrices (P, T, E) suffices for the FEM algorithms; however, sometimes the information stored in these matrices is not enough to answer the following questions:

1. Which elements contain a given mesh point (node)?

2. Which elements share the common edge?

3. If the element's edge is a boundary one?

To answer these question matrices (P, T, E) should be extended by including additional information.

   Let us consider 1st *question.* Connectivity matrix T can be considered as a compact representation of the extended Boolean matrix $T_{ext} \sim (n_p \times n_t)$. Each jth column of the matrix T_{ext} has three nonzero (unity) elements located at positions $T_{1,j}, T_{2,j}$ and $T_{3,j}$. It is not difficult to see that i^{th} row of the matrix T_{ext} has unity on the positions corresponding to the elements having node i as a vertex. Consequently, if we obtain a sparse representation of the matrix T_{ext} (which can be done by introducing vectors of indices i_t and p_t) then we are able to resolve 1st question. This logic is coded through MATLAB function in listing 4.2 which returns sparse format of extended transposed connectivity matrix.

```
1 function [it,pt]=extendT(p,t)
2   np=size(p,2);
3   nt=size(t,2);
4   j =[1:nt;1:nt;1:nt];
5   T_ext=sparse(t(1:3,:),j,1,np,nt); % extended connectivity matrix
6   % coordinate form of T_ext
7   [it,j]=find(T_ext');
8   pt=find(diff([0;j;np+1]))';
9 end
```

Listing 4.2: MATLAB code for generating extended transposed connectivity matrix.

Matrix T_{ext} for the mesh given in figure 4.4 takes the form

$$T_{ext} = \begin{pmatrix} 1 & 0 & 0 & 0 & 0 & 0 \\ 1 & 0 & 0 & 1 & 0 & 0 \\ 0 & 1 & 0 & 0 & 0 & 0 \\ 0 & 1 & 1 & 0 & 0 & 0 \\ 0 & 0 & 0 & 1 & 1 & 0 \\ 0 & 0 & 0 & 0 & 1 & 1 \\ 0 & 0 & 1 & 0 & 0 & 1 \\ 1 & 1 & 1 & 1 & 1 & 1 \end{pmatrix},$$

and the following piece of code

```
[it,pt]=extendT(p,t) ;
for i=1:size(p,2)
    disp([i,it(pt(i):pt(i+1)-1)'])
end
```

results in displaying mesh point index and corresponding triangles having this point as a vertex.

```
i it(pt(i):pt(i+1)-1)
1 1
2 1 4
3 2
4 2 3
5 4 5
6 5 6
7 3 6
8 1 2 3 4 5 6
```

It can be seen that i^{th} mesh node is the vertex of triangles with indices it(pt(i):pt(i+1)-1).

Resolving 2nd and 3rd questions. Let us introduce two complementary matrices E^c and T^c as follows:

- Matrix $E^c \sim (7 \times n_k)$ contains information not only about element's edges belonging to boundary segments (like matrix E), but information about edges of all elements. Here, n_k is the number of edges of all finite elements. Components $E^c_{1,k}$ and $E^c_{2,k}$ are the indices of mesh points belonging to k^{th} edge, $E^c_{3,k}$ and $E^c_{4,k}$ are the values of parameter s for the kth edge provided it belongs to boundary segments and $E^c_{3,k} = E^c_{4,k} = 0$ otherwise, $E^c_{5,k}$ is the segment index, or $E_{5,k}^c = 0$ if the kth edge does not belong to a boundary segment, $E_{6,k}^c$ and $E_{7,k}^c$ are the indices of triangles (elements) from left and right hand sides of the kth edge.

- Matrix $T^c \sim (4 \times n_t)$ contains information regarding indices of edges and subdomain labels. Components $T_{1,j}^c, T_{2,j}^c$ and $T_{3,j}^c$ are the indices of edges of the jth element, $T_{4,j}^c$ is the subdomain label to which the jth element belongs.

Three matrices (P, T^c, E^c) are complementary to (P, T, E) and contain full information regarding the mesh (we further call them complementary $\mathcal{G}_1$ mesh). Complementary $\mathcal{G}_1$ mesh is convenient when elements are treated independently. Following MATLAB function complMeshG1() calculates matrices E^c and T^c

```
1  function [e_c,t_c]=complMeshG1(p,e,t)
2    ne=size(e,2); nt=size(t,2);
3    tt=(t(1:3,:))';
4    ee = [tt(:,[1,2]);tt(:,[2,3]);tt(:,[3,1])]; % not unique edges
5    [ee,~,j]=unique(sort(ee,2),'rows'); % unique mesh edges
6    ee=ee'; j=j'; nee=size(ee,2);
7    mt=1:nt;
8    t_c=[j(mt);j(mt+nt);j(mt+2*nt)]; % elements edges
9    t_c=[t_c;t(4,:)];
10   et=zeros(2,nee); % edge to element connectivity matrix
11   for k=1:nt
12     for j=1:3
13       it=t_c(j,k);
14       if et(1,it)==0, et(1,it)=k;
15       else et(2,it)=k; end
16     end
17   end
18   % sort first 2 rows of e as in ee
19   for i=1:ne
20     if e(1,i)>e(2,i);
21       c=e(1,i); e(1,i)=e(2,i); e(2,i)=c;
22       c=e(3,i); e(3,i)=e(4,i); e(4,i)=c;
23       c=e(6,i); e(6,i)=e(7,i); e(7,i)=c;
24     end
25   end
26   [~,ib,it]=intersect(ee',e(1:2,:)','rows');
```

```
27   % set 3:5 rows of e_c
28   e_c=[ee;zeros(5,nee)];
29   e_c(3:5,ib)=e(3:5,it);
30   % set 6 ,7 rows of e_c
31   for ie=1:nee % ie=edge
32     t1=et(1,ie); t2=et(2,ie); % neighbour triangles
33     i=ee(1,ie); j=ee(2,ie); k=setdiff(tt(t1,:),[i j]);
34     % oriented area of triangle (i,j,k)
35     d12=p(:,j)-p(:,i);
36     d13=p(:,k)-p(:,i);
37     S = d12(1,:).*d13(2,:)-d12(2,:).*d13(1,:);
38     if S>0, e_c(6,ie)=t1; e_c(7,ie)=t2;
39     else e_c(6,ie)=t2; e_c(7,ie)=t1; end
40   end
41 end
```

Listing 4.3: MATLAB code for generating complementary mesh.

Complementary matrices and extended connectivity matrix considered here give more flexibility in manipulating mesh and allow extracting more rich information regarding mesh elements which could be helpful for building refined FEM algorithms.

4.4   BUILDING MESHES IN MATLAB®

Building triangulation in MATLAB for the two-dimensional domain Ω can be performed using function initmesh() which is part of the pde toolbox extension of the MATLAB. For example,

```
[p,e,t]=initmesh ('geometry','Hmax',0.1);
pdemesh (p,e,t); axis equal;
```

Here, function initmesh() builds a mesh defined in the geometry file 'geometry' and command pdemesh() plots the resulting mesh. Parameter 'Hmax' defines the maximum edge value of the mesh triangles. Function initmesh() returns mesh representation in terms of points, edges and connectivity matrices, (p,e,t), described above.

There are two possibilities to define a geometry for the function initmesh():

- using custom-build function;

- using a matrix (when all segments are lines or ellipse segments).

In order a custom-build function defining geometry to be passed to the function initmesh() as a parameter, it should have option to return specific values depending on provided arguments. The function geometryFunction() for geometry encoding discussed in section 4.1 (see listing 4.1) can be passed to the function initmesh() as an argument and it meets interface requirements to be processed as a function argument. For instance, following script generates and plots the mesh shown in figure 4.3 where the geometry is defined by geometryFunction()

```
[p,e,t]=initmesh ('geometryFunction','Hmax',0.1);
pdemesh (p,e,t); axis equal;
```

Another option of geometry definition which can be used by initmesh() function is a matrix where each row must have specific meaning. As an illustration, the geometry shown in figure 4.2 can be encoded through the following matrix definition

```
 1 function g=geometryMatrix
 2    g=[
 3     2    2     2   2 2    2   2 2 2    2 2     4      4      4      4
 4     2 1.25  1.25   2 2 0.75   0 0 0 0.75 0  0.75      1   1.25      1
 5     0 1.25  0.75   2 2 1.25   0 0 2 0.75 2     1   1.25      1   0.75
 6    -1 0.75  0.25  -1 0 0.75  -1 0 1 0.25 0  -0.5  -0.75   -0.5  -0.25
 7    -1 0.25  0.25   0 1 0.75   0 1 1 0.75 0 -0.75   -0.5  -0.25   -0.5
 8     0    3     3   1 3    3   0 0 0    3 3     0      0      0      0
 9     1    2     2   0 0    2   1 3 3    2 1     1      1      1      1
10     0    0     0   0 0    0   0 0 0    0 0     1      1      1      1
11     0    0     0   0 0    0   0 0 0    0 0  -0.5   -0.5   -0.5   -0.5
12     0    0     0   0 0    0   0 0 0    0 0  0.25   0.25   0.25   0.25
13     0    0     0   0 0    0   0 0 0    0 0  0.25   0.25   0.25   0.25
14     0    0     0   0 0    0   0 0 0    0 0     0      0      0      0
15    ] ;
16 end
```

Listing 4.4: Example geometry definition using matrix encoding.

The matrix in listing 4.4 contains 15 columns (number of segments) and each column defines one segment. First element in the column defines type of the segment (2 is a straight line, 4 is an ellipse); 2th and 3th (4th and 5th) elements define x_1 (x_2) coordinates of starting points and endpoints, respectively; 6th and 7th elements specify labels of left and right subdomains. Elements from 8 to 12 are for ellipse specification and can be omitted for straight lines. Elements 8 and 9 define x_1 and

x_2 coordinates of the ellipse centre; elements 10 and 11 are ellipse's semi-axes; 12th element specifies ellipse's rotation angle (anticlockwise in radian).[1] Following piece of code builds the example mesh shown in figure 4.3 using `geometryMatrix` geometry definition.

```
geometry=geometryMatrix;
[p, e, t]=initmesh (geometry,'Hmax',0.1);
pdemesh (p, e, t); axis equal;
```

For better mesh visualization, it is sometimes helpful to display not only mesh as a set of triangles, but labels for individual triangle vertices and edges as well. This would improve mesh illustration and facilitate debugging process for matrix encoding, (p,e,t). Listing 4.5 introduces a function `plotG1mesh()` which uses matrices p, e and t to plot labelled triangulation. Function `plotG1mesh()` has the following parameters: p,e,t are the matrices defining the mesh, opt is a Boolean vector for option specification (opt(1)==1 – outputs point labels, opt(2)==1 – outputs boundary edge labels, opt(3)==1 – outputs triangle labels), and fs is a font size.

```
 1  function h=plotG1mesh (p,e,t,opt,fs)
 2    h=figure;
 3    pdemesh(p,e,t); axis equal; hold on
 4    plot(p(1,:),p(2,:),'.k','MarkerSize',8)
 5    xlabel('x_1'), ylabel('x_2')
 6    if opt(1)==1 % output point labels
 7      for k=1:size(p,2)
 8        text(p(1,k),p(2,k),[' ' int2str(k)], ...
 9            'FontSize',fs,'Color','r');
10      end
11    end
12    if opt(2)==1 % output boundary labels
13      for k=1:size(e,2)
14        i=e(1,k); j=e(2,k);
15        x1=(p(1,i)+p(1,j))/2; x2=(p(2,i)+p(2,j))/2;
16        text(x1,x2+0.05,[' ' int2str(e(5,k))], ...
17            'FontSize',fs,'Color','k');
18      end
19    end
20    if opt(3)==1 % output element labels
21      for k=1:size(t,2)
```

---

[1]The matrix has only 7 rows if all segments are straight lines.

```
22      I=t(1:3,k); x1x2=sum(p(:,I)')/3;
23      text(x1x2(1),x1x2(2),int2str(k),'FontSize',fs,'Color','b');
24    end
25  end
26 end
```

Listing 4.5: MATLAB code which plots labelled mesh.

The L-shaped labelled geometry shown in figure 4.4 is generated through the script

```
[p, e, t]=initmesh ('lshapeg','Hmax', inf);
plotG1mesh (p, e, t, [1, 1, 1], 12);
```

File lshapeg.m for this geometry is provided in the MATLAB pde toolbox (parameter inf indicates building rough mesh). Figure 4.4 shows triangulation with given labels of triangles, edges and vertices and function plotG1mesh() plots this triangulation.

Similarly, complementary mesh discussed in section 4.3 can also be labelled with corresponding indices using following script

```
[p,e,t]=initmesh('lshapeg','hmax',inf);
[e_c,t_c]=complMeshG1(p,e,t);
plotComplMeshG1(p,e,t,e_c,12);
```

where the function plotComplMeshG1() plots complementary mesh and has a form given in listing 4.6

```
1 function h=plotComplMeshG1(p,e,t,e_c,fs)
2   h=figure; pdemesh(p,e,t); axis equal; hold on
3   plot(p(1,:),p(2,:),'.k','MarkerSize',8);
4   xlabel('x_1'), ylabel('x_2')
5   for i = 1:size(p,2)
6     text(p(1,i),p(2,i),[' ' int2str(i)], ...
7         'FontSize',fs—1,'Color','k');
8   end
9   for it=1:size(e_c,2) % edge labels
10    i=e_c(1,it); j=e_c(2,it);
11    text((p(1,i)+p(1,j))/2—0.06,(p(2,i)+p(2,j))/2, ...
12      [' ' int2str(it)],'FontSize',fs—2,'Color','r');
13    if e_c(5,it);
14      text((p(1,i)+p(1,j))/2+0.05,(p(2,i)+p(2,j))/2+0.05, ...
15        [' ' int2str(e_c(5,it))],'FontSize',fs—2,'Color','k');
16    end
```

```
17    end
18    for i=1:size(t,2) % element labels
19      I=t(1:3,i);
20      text(sum(p(1,I))/3,sum(p(2,I))/3,int2str(i), ...
21          'FontSize',fs-1,'Color','b');
22    end
23 end
```

Listing 4.6: MATLAB code which plots labelled complementary mesh.

As a result, we obtain figure 4.5 with given indices for mesh points, triangle elements and edges. Boundary edges are marked with two numbers: index of the edge and index of the boundary segment.

4.5 BUILDING MESHES IN PYTHON

Above, we have considered how to define a domain geometry and build a mesh using MATLAB tools. Now, we explain how to perform mesh building in Python. Again, the approach is to use third-party library

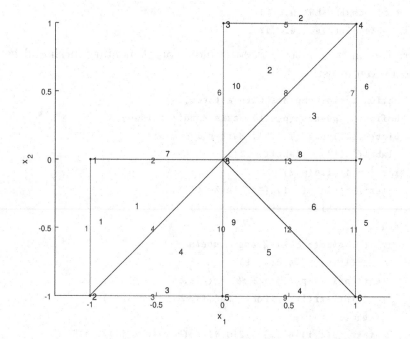

FIGURE 4.5: Complementary mesh.

for mesh processing rather than discussing particular mesh building algorithm. There is a powerful mesh generation tool, named Gmsh, having its own scripting language allowing creating complex geometries. The library PyGmsh provides a Python interface for the Gmsh scripting language so one can create geometry and generate meshes more easily with excess to Python's features. After installing PyGmsh, the default Gmsh kernel with basic geometry construction functions can be used.

Module PyGmsh has a class Geometry to represent a geometry in the following way: first, it is to be defined points (using the method add_point), then lines (using add_line, add_circle, add_circle_arc, add_line_loop, etc.), and finally surfaces (using add_plane_surface method). These geometrical entities are assigned identification numbers when they are created. Groups of geometrical entities can be defined (using add_physical_line or add_physical_surface methods) and are called physical entities with unique identification numbers. The purpose of these physical entities is to assemble geometrical entities into larger groups so that they can be referred to by the mesh module as single entities. In terms of domain geometry definitions, physical entities can be viewed as labelled boundaries or subdomains on which boundary conditions or coefficients are defined. As an example, script in listing 4.7 defines geometry illustrated in figures 4.2 and 4.3. Function geometry() has a parameter h – mesh triangular size, and returns two structures: geometry object, geom, and labels of geometry elements, geom_labels.

```python
1  import pygmsh
2  import numpy
3  def geometry(h):
4      geom = pygmsh.built_in.Geometry() # initialize geometry object
5      #rectangular #1
6      rect1_p1=geom.add_point([0.0, -1.0, 0.0], h)
7      rect1_p2=geom.add_point([0.0, 0.0, 0.0], h)
8      rect1_p3=geom.add_point([2.0, 0.0, 0.0], h)
9      rect1_p4=geom.add_point([2.0, -1.0, 0.0], h)
10     rect1_l1=geom.add_line(rect1_p1, rect1_p2)
11     rect1_l2=geom.add_line(rect1_p2, rect1_p3)
12     rect1_l3=geom.add_line(rect1_p3, rect1_p4)
13     rect1_l4=geom.add_line(rect1_p4, rect1_p1)
14     # circular hole
15     cirlcle_p0=geom.add_point([1.0, -0.5, 0.0], h) #center
16     cirlcle_p1=geom.add_point([0.75, -0.5, 0.0], h)
```

```
17    cirlcle_p2=geom.add_point([1.0, −0.75, 0.0], h)
18    cirlcle_p3=geom.add_point([1.25, −0.5, 0.0], h)
19    cirlcle_p4=geom.add_point([1.0, −0.25, 0.0], h)
20    arc_1=geom.add_circle_arc(cirlcle_p1, cirlcle_p0, cirlcle_p2)
21    arc_2=geom.add_circle_arc(cirlcle_p2, cirlcle_p0, cirlcle_p3)
22    arc_3=geom.add_circle_arc(cirlcle_p3, cirlcle_p0, cirlcle_p4)
23    arc_4=geom.add_circle_arc(cirlcle_p4, cirlcle_p0, cirlcle_p1)
24    rect1_line=geom.add_line_loop([rect1_l1,rect1_l2, \
25                             rect1_l3,rect1_l4, \
26                             arc_1,arc_2,arc_3,arc_4])
27    #rectangle #2
28    rect2_p1=geom.add_point([0.75, 0.25, 0.0], h)
29    rect2_p2=geom.add_point([0.75, 0.75, 0.0], h)
30    rect2_p3=geom.add_point([1.25, 0.75, 0.0], h)
31    rect2_p4=geom.add_point([1.25, 0.25, 0.0], h)
32    rect2_l1=geom.add_line(rect2_p1, rect2_p2)
33    rect2_l2=geom.add_line(rect2_p2, rect2_p3)
34    rect2_l3=geom.add_line(rect2_p3, rect2_p4)
35    rect2_l4=geom.add_line(rect2_p4, rect2_p1)
36    rect2_line=geom.add_line_loop([rect2_l1,rect2_l2, \
37                             rect2_l3,rect2_l4])
38    #rectangle #3
39    rect3_p1=geom.add_point([0.0, 0.0, 0.0], h)
40    rect3_p2=geom.add_point([0.0, 1.0, 0.0], h)
41    rect3_p3=geom.add_point([2.0, 1.0, 0.0], h)
42    rect3_p4=geom.add_point([2.0, 0.0, 0.0], h)
43    rect3_l1=geom.add_line(rect3_p1, rect3_p2)
44    rect3_l2=geom.add_line(rect3_p2, rect3_p3)
45    rect3_l3=geom.add_line(rect3_p3, rect3_p4)
46    rect3_l4=geom.add_line(rect3_p4, rect3_p1)
47    rect3_line=geom.add_line_loop([rect3_l1,rect3_l2, \
48                             rect3_l3,rect3_l4,rect2_l1, \
49                             rect2_l2,rect2_l3,rect2_l4])
50    #create surfaces
51    rect1_surf=geom.add_plane_surface(rect1_line)
52    rect3_surf=geom.add_plane_surface(rect3_line)
53    rect2_surf=geom.add_plane_surface(rect2_line)
54    # define labels for geometry elements:
55    geom_labels={}
56    group_id=geom._TAKEN_PHYSICALGROUP_IDS
```

```
57    # domain labels
58    ph_rect1_surf=geom.add_physical_surface(rect1_surf,label="1")
59    geom_labels[group_id[-1]]="1"
60    ph_rect2_surf=geom.add_physical_surface(rect2_surf,label="2")
61    geom_labels[group_id[-1]]="2"
62    ph_rect3_surf=geom.add_physical_surface(rect3_surf,label="3")
63    geom_labels[group_id[-1]]="3"
64    # boundary labels
65    ph_rect1_l1=geom.add_physical_line(rect1_l1,label="7:L0:R1")
66    geom_labels[group_id[-1]]="7:L0:R1"
67    ph_rect1_l2=geom.add_physical_line(rect1_l2,label="11:L3:R1")
68    geom_labels[group_id[-1]]="11:L3:R1"
69    ph_rect1_l3=geom.add_physical_line(rect1_l3,label="4:L0:R1")
70    geom_labels[group_id[-1]]="4:L0:R1"
71    ph_rect1_l4=geom.add_physical_line(rect1_l4,label="1:L0:R1")
72    geom_labels[group_id[-1]]="1:L0:R1"
73    ph_rect2_l1=geom.add_physical_line(rect2_l1,label="10:L3:R2")
74    geom_labels[group_id[-1]]="10:L3:R2"
75    ph_rect2_l2=geom.add_physical_line(rect2_l2,label="6:L3:R2")
76    geom_labels[group_id[-1]]="6:L3:R2"
77    ph_rect2_l3=geom.add_physical_line(rect2_l3,label="2:L3:R2")
78    geom_labels[group_id[-1]]="2:L3:R2"
79    ph_rect2_l4=geom.add_physical_line(rect2_l4,label="3:L3:R2")
80    geom_labels[group_id[-1]]="3:L3:R2"
81    ph_rect3_l1=geom.add_physical_line(rect3_l1,label="8:L0:R3")
82    geom_labels[group_id[-1]]="8:L0:R3"
83    ph_rect3_l2=geom.add_physical_line(rect3_l2,label="9:L0:R3")
84    geom_labels[group_id[-1]]="9:L0:R3"
85    ph_rect3_l3=geom.add_physical_line(rect3_l3,label="5:L0:R3")
86    geom_labels[group_id[-1]]="5:L0:R3"
87    ph_arc_1=geom.add_physical_line(arc_1,label="12:L0:R1")
88    geom_labels[group_id[-1]]="12:L0:R1"
89    ph_arc_2=geom.add_physical_line(arc_2,label="13:L0:R1")
90    geom_labels[group_id[-1]]="13:L0:R1"
91    ph_arc_3=geom.add_physical_line(arc_3,label="14:L0:R1")
92    geom_labels[group_id[-1]]="14:L0:R1"
93    ph_arc_4=geom.add_physical_line(arc_4,label="15:L0:R1")
94    geom_labels[group_id[-1]]="15:L0:R1"
```

```
95     return geom,geom_labels
```

Listing 4.7: Python code for defining example geometry.

Let us make some comments regarding defining labels for the geometry elements. Subdomains are labelled by numbers, but for labelling a boundary (see lines 65 –94) we use group of colon-separated three labels: a number, letter L with a number and letter R with a number. Here, first number is the boundary label and subsequent numbers are subdomain labels from left (letter L) and right (letter R) hand sides of the boundary. These labels are assigned to physical groups of the geometry and kept in the dictionary geom_labels which maps identification number of each physical group to the specified label. Group identification numbers are generated automatically and kept in the list _TAKEN_PHYSICALGROUP_IDS of the class Geometry.

The next step is to generate a mesh and retrieve triangulation data for p,t,e matrices (see section 4.2 for detailed explanation of data structure p,t,e). Function generateMesh() in listing 4.8 creates matrices of point coordinates, p, connectivity, t and edges e which have same structure as described in section 4.2. It has two parameters: geom – geometry object defining geometry structure, and geom_labels – a structure containing labels for geometry elements (these structures are created by geometry() function), and returns matrices (p,t,e).

```
1  import pygmsh
2  import numpy
3  def generateMesh(geom,geom_labels):
4      # generate mesh
5      points, cells, point_data, cell_data, field_data= \
6                                  pygmsh.generate_mesh(geom)
7      #extract p,t,e matrices
8      p=points[:,0:2]
9      t=numpy.zeros((cells["triangle"].shape[0],4))
10     e=numpy.zeros((cells["line"].shape[0],7))
11     t[:,0:3]=cells["triangle"]
12     e[:,0:2]=cells["line"]
13     for label in field_data:
14         elemnt_index=field_data[label][0]
15         group_index=field_data[label][1]
16         names=list(cell_data.keys())
17         if names[group_index-1]=="triangle":# domain labels
18             t[cell_data["triangle"] \
```

```
19                    ["gmsh:physical"]==elemnt_index ,3]= \
20                    int(geom_labels[elemnt_index])
21         elif names[group_index —1]=="line":# boundary labels
22            label=None
23            left_label=None
24            right_label=None
25            for item in geom_labels[elemnt_index].split(":"):
26                if item[0]=="L": left_label=int(item[1:])
27                elif item[0]=="R": right_label=int(item[1:])
28                else: label=int(item)
29            if label==None or \
30                left_label==None or \
31                right_label==None:
32                print('error, wrong geometry boundary labelling')
33            index=cell_data["line"]["gmsh:physical"]==elemnt_index
34            e[index,4]=label
35            e[index,5]=left_label
36            e[index,6]=right_label
37    return p,t.astype(int),e.astype(int)
```

Listing 4.8: Python code for building a mesh from given geometry.

It is worth noting that as we will see later, elements of the edge matrix, e, corresponding to geometry parameter values generated by the MATLAB initmesh() function are not in use in FEM framework, as such, to be consistent, in listing 4.8, we keep the same number of rows in matrix e but not filling those for parameter values of edges.

As a result, the mesh illustrated in figure 4.3 can be generated through running the script

```
geom,geom_labels=geometry(0.5)
p,t,e = generateMesh(geom,geom_labels)
```

CHAPTER 5

Programming Two-Dimensional Finite Element Method

I N this chapter we will consider five different methods of assembling global stiffness matrix and global force vector with evaluation of computational cost, efficiency and performance of each method in details. First, we will be focussed on matrices assembling technique, not on calculating matrices elements, and assuming generic local stiffness matrix to be a unity matrix. Further, we will consider methods for calculating elements of local stiffness matrices and local forcing vectors. Also, it will be shown how to incorporate boundary conditions into FEM. Description given in this chapter will involve some mathematical analysis which is important to understand computational techniques. Numerical methods for calculating integrals, function transformations and operating with basis functions will be used to deduce algorithms for calculating local FEM matrices.

5.1 ASSEMBLING GLOBAL STIFFNESS MATRIX

Let us consider assembling global stiffness matrix for $\mathcal{G}_1$ triangular mesh corresponding to a square domain. Triangulation is represented in terms of matrices (p,e,t). It is important to mention that t(1:3,it) stores global node indices for the element having index it. Local stiffness matrices could be assembled into the global one through the following five methods.

DOI: 10.1201/9781003265979-5

1st *algorithm (inefficient)*. This algorithm is based on the definition of global stiffness matrix and assembles the stiffness matrix element by element, but it is computationally inefficient and not recommended for practical use in MATLAB.

```matlab
1  function A=stiffnessAssemb_1(p,t)
2    np=size(p,2) ;
3    nt=size(t,2) ;
4    ntep=3; % number of triangular element points
5    A=sparse(np,np) ; % global stiffness matrix
6    Al=ones(ntep,ntep) ; % local stiffness matrix
7    for k=1:nt
8      I=t(1:ntep,k) ; % global point indices on element k
9      for i=1:ntep
10       for j=1:ntep
11         A(I(i),I(j))=A(I(i),I(j))+Al(i,j) ;
12       end
13     end
14   end
15 end
```

Listing 5.1: First version of the code for assembling stiffness matrix which uses element by element strategy.

2nd *algorithm (partly optimized)*. This is a vectorized version of the previous algorithm and it is suitable for meshes with small number of elements.

```matlab
1  function A=stiffnessAssemb_2(p,t)
2    np=size(p,2) ;
3    nt=size(t,2) ;
4    ntep=3;
5    A=sparse(np,np) ; % global stiffness matrix
6    Al=ones(ntep,ntep) ; % local stiffness matrix
7    for k=1:nt
8      I=t(1:ntep,k) ;
9      A(I,I)=A(I,I)+Al ; % assembling matrix
10   end
11 end
```

Listing 5.2: Second version of the code for assembling stiffness matrix which uses partly vectorized approach.

3^d *algorithm (optimized)*. This algorithm is based on assembling the stiffness matrix in the coordinate form.

```
1  function A=stiffnessAssemb_3(p,t)
2    np=size(p,2) ;
3    nt=size(t,2) ;
4    ntep=3;
5    ntep2=ntep^2; % number of local stiffness matrix elements
6    m=ntep2*nt ; % number of elemnts of all local matrices
7    % coordinates representation of matrix A
8    i=zeros(m,1) ;
9    j=zeros(m,1) ;
10   v=ones(m,1) ;
11   Al=rand(ntep,ntep) ; % local stiffness matrix
12   % assembling matrix A in coordinate form (i,j,v)
13   for k=1:nt
14     I=t(1:ntep,k) ;
15     il=repmat(I(:),1,ntep) ;
16     jl=il';
17     m=ntep2*(k-1)+(1:ntep2) ;
18     i(m)=il(:) ;
19     j(m)=jl(:) ;
20     v(m)=Al(:) ;
21   end
22   A=sparse(i,j,v,np,np) ;
23 end
```

Listing 5.3: Third version of the code for assembling stiffness matrix which uses coordinate form of sparse matrix.

Some comments are to be mentioned here. For any matrix B, indexing like B(:) transforms the matrix into a column of the length numel(B). Element Al(i,j) should be added to the corresponding positions of matrix A. As such, instructions i(m)=il(:) ; j(m)=jl(:) ; v(m)=Al(:) transform matrices il, jl and Al into columns and assign them to the corresponding elements of the vectors i, j and v.

4th *algorithm (most efficient)*. This algorithm is based on fully vectorized form of assembling global stiffness matrix

```
1  function A=stiffnessAssemb_4(p,t)
2    np=size(p,2) ;
3    nt=size(t,2) ;
```

```
4    %indices of local points
5    k1=t(1,:) ;
6    k2=t(2,:) ;
7    k3=t(3,:) ;
8    % aij(m)=Al(i,j) on element m
9    a12=ones(nt,1) ; A=sparse(k1,k2,a12,np,np) ;
10   a13=ones(nt,1) ; A=A+sparse(k1,k3,a13,np,np) ;
11   a23=ones(nt,1) ; A=A+sparse(k2,k3,a23,np,np) ;
12   % replace next 3 lines by A=A+A.' for symmetric A
13   a21=ones(nt,1) ; A=A+sparse(k2,k1,a21,np,np) ;
14   a31=ones(nt,1) ; A=A+sparse(k3,k1,a31,np,np) ;
15   a32=ones(nt,1) ; A=A+sparse(k3,k2,a32,np,np) ;
16   a11=ones(nt,1) ; A=A+sparse(k1,k1,a11,np,np) ;
17   a22=ones(nt,1) ; A=A+sparse(k2,k2,a22,np,np) ;
18   a33=ones(nt,1) ; A=A+sparse(k3,k3,a33,np,np) ;
19   end
```

Listing 5.4: Fourth version of the code for assembling stiffness matrix which uses fully vectorized approach.

Function stiffnessAssemb_4() is loop free and operates only with vectors. For the symmetric matrix A the algorithm could be more memory efficient (in this case some of aij can be replaced by a single vector and A=A+A.'). We can avoid vectorization of the local stiffness matrices in function stiffnessAssemb_4() and save some memory. For example, the following version of assembling algorithm is comparable in terms of computational cost with the previous one

```
1    function A=stiffnessAssemb_5(p,t)
2    np=size(p,2) ;
3    nt=size(t,2) ;
4    ntep=3;
5    %indices of local points
6    k1=t(1,:) ;
7    k2=t(2,:) ;
8    k3=t(3,:) ;
9    Al=ones(nt,ntep,ntep) ; % all local stiffness matrices
10   A=sparse(k1,k2,Al(:,1,2),np,np) ;
11   A=A+sparse(k1,k3,Al(:,1,3),np,np) ;
12   A=A+sparse(k2,k3,Al(:,2,3),np,np) ;
13   % replace next 3 lines by A=A+A.'for symmetric A
14   A=A+sparse(k2,k1,Al(:,2,1),np,np) ;
```

```
15   A=A+sparse(k3,k1,Al(:,3,1),np,np) ;
16   A=A+sparse(k3,k2,Al(:,3,2),np,np) ;
17   A=A+sparse(k1,k1,Al(:,1,1),np,np) ;
18   A=A+sparse(k2,k2,Al(:,2,2),np,np) ;
19   A=A+sparse(k3,k3,Al(:,3,3),np,np) ;
20 end
```

Listing 5.5: Fifth version of the code for assembling stiffness matrix which uses fully vectorized approach with improved memory usage efficiency.

Let us compare computational efficiency of all assembling algorithms with the help of the function testAssembStiffness() in listing 5.6.

```
1  function testAssembStiffness
2    clc
3    % assembling time
4    time_1=[] ; time_2=[] ; time_3=[] ; time_4=[] ; time_5=[] ;
5    nt=[] ; np=[] ;
6    for nx=[10 50 100]
7      % set a mesh of rectangular domain
8      [p,e,t] =poimesh('squareg',nx,nx) ;
9      np=[np size(p,2)] ;
10     nt=[nt size(t,2)] ;
11     tic , stiffnessAssemb_1(p,t) ; time_1=[time_1 toc] ;
12     tic , stiffnessAssemb_2(p,t) ; time_2=[time_2 toc] ;
13     tic , stiffnessAssemb_3(p,t) ; time_3=[time_3 toc] ;
14     tic , stiffnessAssemb_4(p,t) ; time_4=[time_4 toc] ;
15     tic , stiffnessAssemb_5(p,t) ; time_5=[time_5 toc] ;
16   end
17   disp('np  nt  ver_1   ver_2   ver_3   ver_4   ver_5')
18   disp([np' nt' time_1' time_2' time_3' time_4' time_5'])
19 end
```

Listing 5.6: Script for comparing efficiency of various stiffness matrix assembling strategies.

Table 5.1 summarizes results of the computational efficiency test. We can conclude that the CPU time for the first and second versions of the function stiffnessAssemb() is proportional to the square of nodes number while CPU time for the third, fourth and fifth versions of the assembling algorithm are practically proportion to the number of mesh nodes. Algorithms of ver_4 and ver_5 are faster than ver_3 since they do

TABLE 5.1: Summary of the computational efficiency test. Last five columns contain CPU time for corresponding version of the stiffness matrix assembling algorithm.

np	nt	ver_1	ver_2	ver_3	ver_4	ver_5
121	200	0.04883	0.00586	0.05566	0.0	0.0
2601	5000	1.85449	0.66308	1.36817	0.00293	0.00293
10201	20000	17.21289	8.50488	5.65820	0.01270	0.01367

not contain loops. As a result, it is recommended to use vectorized form of the global matrix assembling algorithm.

For arbitrary finite elements (not necessarily triangle) we recommend using generic function shown in listing 5.7 for assembling global stiffness matrix (provided there is a function calcLocalStiffness() to compute a local stiffness matrix). Here, p is a matrix of points, t is a matrix of elements, and ntep is a number of element's points.

```
1  function A=stiffnessAssembGeneric(ntep,p,t)
2    np=size(p,2) ;
3    nt=size(t,2) ;
4    Al=zeros(ntep,ntep,nt) ; % all local stiffness matrices
5    for k=1:nt
6      Al(k,:,:)=calcLocalStiffness(k,p,t) ; % local stiffness matrix
7    end
8    A=sparse(np,np) ; % global matrix
9    for i=1:ntep
10     for j=1:ntep
11       A=A+sparse(t(i,:),t(j,:),Al(:,i,j),np,np) ;
12     end
13   end
14 end
```

Listing 5.7: Version of the code for assembling stiffness matrix with finite elements of arbitrary shape.

5.2   ASSEMBLING GLOBAL FORCING VECTOR

Assembling global forcing vector, F, is performed in a similar to global stiffness matrix way. Three methods of assembling could be used depending on the possibility to vectorize calculations. Functions below implement these methods.

```
1 function F=forcingAssemb_1(p,t)
2   np=size(p,2) ; nt=size(t,2) ;
3   k1=t(1,:) ; k2=t(2,:) ; k3=t(3,:) ;
4   % Fl - local forcing vector
5   %fj(m)= Fl(j) on element m,
6   f1=ones(nt,1) ; F=sparse(k1,1,f1,np,1) ;
7   f2=ones(nt,1) ; F=F+sparse(k2,1,f2,np,1) ;
8   f3=ones(nt,1) ; F=F+sparse(k3,1,f3,np,1) ;
9 end
```

Listing 5.8: Version of the code for assembling forcing vector which uses fully vectorized approach.

```
1 function F=forcingAssemb_2(p,t)
2   np=size(p,2) ; nt=size(t,2) ;
3   k1=t(1,:) ; k2=t(2,:) ; k3=t(3,:) ;
4   Fl=ones(nt,3) ; % all local forcing vectors
5   F=sparse(k1,1,Fl(:,1),np,1);
6   F=F+sparse(k2,1,Fl(:,2),np,1) ;
7   F=F+sparse(k3,1,Fl(:,3),np,1) ;
8 end
```

Listing 5.9: Version of the code for assembling forcing vector which uses fully vectorized approach with improved memory usage efficiency.

```
1 function F=forcingAssemb_3(p,t)
2   np=size(p,2); nt=size(t,2) ; ntep=3;
3   F=zeros(np,1) ;
4   for k=1:nt
5     I=t(1:ntep,k) ;
6     Fl=ones(ntep,1) ;
7     F(I)=F(I)+Fl ;
8   end
9 end
```

Listing 5.10: Version of the code for assembling forcing vector with finite elements of arbitrary shape.

To test performance let us run the following script

```
% testing forcing vector assembling functions
time_1=[] ; time_2=[] ; time_3=[] ;% assembling times
nt=[] ; np=[] ;
```

```
for nx=[10 50 100]
    % set a mesh on the domain
    [p,~,t]=poimesh('squareg',nx,nx) ;
    np=[np size(p,2)] ; nt=[nt size(t,2)] ;
    tic ; F=forcingAssemb_1(p,t) ; time_1=[time_1 toc] ;
    tic ; F=forcingAssemb_2(p,t) ; time_2=[time_2 toc] ;
    tic ; F=forcingAssemb_3(p,t) ; time_3=[time_3 toc] ;
end
disp('np nt ver_1 ver_2 ver_3')
disp([np' nt' time_1' time_2' time_3'])
```

which results in table 5.2. It follows from table 5.2 that computational cost for the third method five times higher compared with first or second method.

5.3 CALCULATING LOCAL STIFFNESS MATRICES

Let us consider calculating local stiffness matrix corresponding to a triangular element τ (here we drop the index l implying that τ could be any element from τ_l defined above). Components of the local matrix $\bar{A}_{\alpha\beta}$ with indices $\alpha, \beta = 1, 2, 3$ are defined as

$$\bar{A}_{\alpha\beta} = \int_\tau \left(c\nabla\varphi_{i_\beta} \cdot \nabla\varphi_{i_\alpha} + \boldsymbol{b} \cdot \nabla\varphi_{i_\beta}\varphi_{i_\alpha} + a\varphi_{i_\beta}\varphi_{i_\alpha} \right) d\boldsymbol{x}. \qquad (5.1)$$

Here, i_1, i_2 and i_3 are the vertices indices of the element τ, and φ_{i_1}, φ_{i_2} and φ_{i_3} are the corresponding basis functions (note that (i_1, i_2, i_3) is a column of matrix t). Figure 5.1 shows the element τ with notation. To compute integral (5.1) we need expressions for the basis functions and their gradients along with numerical method for integral calculation.

TABLE 5.2: Summary of the computational efficiency test for the algorithm of assembling global forcing vector. Last three columns contain CPU time for the corresponding version of the algorithm.

np	nt	ver_1	ver_2	ver_3
121	200	0.0	0.00098	0.00879
2601	5000	0.00098	0.00098	0.13281
10201	20000	0.00293	0.00195	0.52637

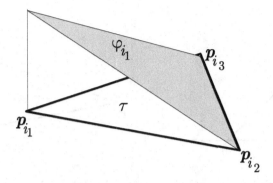

FIGURE 5.1: Triangular element τ and basis function φ_{i_1}.

We consider two possible methods of calculating integral (5.1). First method is based on the assumption that coefficients $c(\boldsymbol{x})$, $b(\boldsymbol{x})$ and $a(\boldsymbol{x})$ are constant over the element τ and equal to the values computed at the element's centre of mass, $\boldsymbol{p}_\tau = (\boldsymbol{p}_{i_1} + \boldsymbol{p}_{i_2} + \boldsymbol{p}_{i_3})/3$. In this case the integral is computed analytically (provided expressions for the basis functions are known) through the formula

$$\bar{A}_{\alpha\beta} = \int_\tau \left(c(\boldsymbol{p}_\tau)\nabla\varphi_{i_\beta} \cdot \nabla\varphi_{i_\alpha} + \boldsymbol{b}(\boldsymbol{p}_\tau) \cdot \nabla\varphi_{i_\beta}\varphi_{i_\alpha} + a(\boldsymbol{p}_\tau)\varphi_{i_\beta}\varphi_{i_\alpha} \right) d\boldsymbol{x}.$$

(5.2)

Second method makes use of the quadrature formula with one node which is faithful representation for linear polynomial integrand

$$\bar{A}_{\alpha\beta} = |\tau|\left(c\nabla\varphi_{i_\beta} \cdot \nabla\varphi_{i_\alpha} + \boldsymbol{b} \cdot \nabla\varphi_{i_\beta}\varphi_{i_\alpha} + a\varphi_{i_\beta}\varphi_{i_\alpha} \right)(\boldsymbol{p}_\tau), \qquad (5.3)$$

where $|\tau|$ is the area of the element τ.

Next, we consider representation of the basis functions on the element τ. To obtain expressions for the basis functions we use an approach which is equally applicable to any type of triangular elements with straight line edges (for example, triangular elements with six nodes or $\mathcal{G}_2$ elements). The main idea is to consider a canonical element $\hat{\tau}$ in the coordinate system $\hat{\boldsymbol{x}} = (\hat{x}_1, \hat{x}_2)$ on which basis functions take simple known form. Then the basis functions for the element τ are derived through coordinate transformation. To illustrate this approach, let us consider two elements shown in figure 5.2, where the canonical element has vertices coordinates $\hat{\boldsymbol{p}}_1 = (0,0)$, $\hat{\boldsymbol{p}}_2 = (1,0)$, $\hat{\boldsymbol{p}}_3 = (0,1)$ and an arbitrary

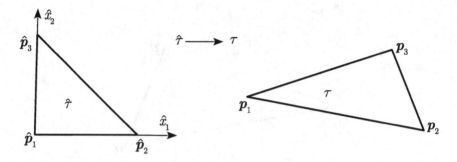

FIGURE 5.2: Canonical element $\hat{\tau}$ and original $\mathcal{G}_1$ mesh element τ.

three-nodes element τ has vertices $\boldsymbol{p}_1$, $\boldsymbol{p}_2$ and $\boldsymbol{p}_3$. In this case, the following functions

$$\hat{\varphi}_1(\hat{\boldsymbol{x}}) = 1 - \hat{x}_1\hat{x}_2, \quad \hat{\varphi}_2(\hat{\boldsymbol{x}}) = \hat{x}_1, \quad \hat{\varphi}_3(\hat{\boldsymbol{x}}) = \hat{x}_2$$

represent basis functions on element $\hat{\tau}$, and it is clear that $\hat{\varphi}_i(\hat{\boldsymbol{p}}_j) = \delta_{ij}$. Transformation from $\hat{\tau}$ to τ is given by

$$\boldsymbol{x} = \boldsymbol{p}_1\hat{\varphi}_1 + \boldsymbol{p}_2\hat{\varphi}_1 + \boldsymbol{p}_3\hat{\varphi}_1 = \boldsymbol{p}_1 + \hat{x}_1(\boldsymbol{p}_2 - \boldsymbol{p}_1) + \hat{x}_2(\boldsymbol{p}_3 - \boldsymbol{p}_1). \quad (5.4)$$

Indeed, this is an affine transformation which maps vertices of $\hat{\tau}$ to the vertices of τ with preserving orientation of the canonical element and $(0,0) \to \boldsymbol{p}_1$, $(1,0) \to \boldsymbol{p}_2$, $(0,1) \to \boldsymbol{p}_3$. After introducing notation

$$B_\tau = \begin{pmatrix} (\boldsymbol{p}_2 - \boldsymbol{p}_1)_1 & (\boldsymbol{p}_3 - \boldsymbol{p}_1)_1 \\ (\boldsymbol{p}_2 - \boldsymbol{p}_1)_2 & (\boldsymbol{p}_3 - \boldsymbol{p}_1)_2 \end{pmatrix}, \quad (5.5)$$

transformation can be rewritten in matrix form

$$\boldsymbol{x} = B_\tau\hat{\boldsymbol{x}} + \boldsymbol{p}_1 \quad : \quad \hat{\tau} \to \tau \quad (5.6)$$

with Jacobian determinant $J_\tau = \det B_\tau > 0$. As a result, the area of element τ is defined as

$$|\tau| = \int_\tau d\boldsymbol{x} = \int_\tau \det B_\tau d\hat{\boldsymbol{x}} = |\hat{\tau}| \det B_\tau = \frac{1}{2}J_\tau.$$

Inverse transformation has a form of $\hat{\boldsymbol{x}} = B_\tau^{-1}(\boldsymbol{x} - \boldsymbol{p}_1)$ with

$$B_\tau^{-1} = \frac{1}{\det B_\tau} \begin{pmatrix} (\boldsymbol{p}_3 - \boldsymbol{p}_1)_2 & -(\boldsymbol{p}_3 - \boldsymbol{p}_1)_1 \\ -(\boldsymbol{p}_2 - \boldsymbol{p}_1)_2 & (\boldsymbol{p}_2 - \boldsymbol{p}_1)_1 \end{pmatrix} = \begin{pmatrix} \tilde{d} & -\tilde{b} \\ -\tilde{c} & \tilde{a} \end{pmatrix}.$$

Consequently,

$$\hat{x}_1 = \tilde{d}\,(x_1 - (\boldsymbol{p}_1)_1) - \tilde{b}\,(x_2 - (\boldsymbol{p}_1)_2)\,,$$
$$\hat{x}_2 = -\tilde{c}\,(x_1 - (\boldsymbol{p}_1)_1) + \tilde{a}\,(x_2 - (\boldsymbol{p}_1)_2)\,.$$

Basis functions $\varphi_i(\boldsymbol{x})$ on the element τ are obtained from the basis functions on the element $\hat{\tau}$ through the transformation $\varphi_i(\boldsymbol{x}) = \hat{\varphi}_i(B_\tau^{-1}(\boldsymbol{x} - \boldsymbol{p}_1))$. As a result,

$$\varphi_1(\boldsymbol{x}) = 1 - \varphi_2(\boldsymbol{x}) - \varphi_3(\boldsymbol{x})\,,$$
$$\varphi_2(\boldsymbol{x}) = \tilde{d}\,(x_1 - (\boldsymbol{p}_1)_1) - \tilde{b}\,(x_2 - (\boldsymbol{p}_1)_2)\,,$$
$$\varphi_3(\boldsymbol{x}) = -\tilde{c}\,(x_1 - (\boldsymbol{p}_1)_1) + \tilde{a}\,(x_2 - (\boldsymbol{p}_1)_2)\,.$$

Respectively, gradients of the basis functions have the form

$$\nabla\varphi_1 = \begin{pmatrix} \tilde{c} - \tilde{d} \\ \tilde{b} - \tilde{a} \end{pmatrix}, \quad \nabla\varphi_2 = \begin{pmatrix} \tilde{d} \\ -\tilde{b} \end{pmatrix}, \quad \nabla\varphi_3 = \begin{pmatrix} -\tilde{c} \\ \tilde{a} \end{pmatrix}.$$

It is worth noting that $\varphi_i(\boldsymbol{p}_\tau) = 1/3$, $i = 1, 2, 3$, and for gradients we have

$$\nabla\varphi_i(\boldsymbol{x}) = B_\tau^{-1\,T}\hat{\nabla}\hat{\varphi}_i(\hat{\boldsymbol{x}}).$$

Also, the following expressions are valid for the elements τ of any type

$$\sum_{i=1}^{m_\tau} \varphi_i(\boldsymbol{x}) = 1, \quad \sum_{i=1}^{m_\tau} \nabla\varphi_i(\boldsymbol{x}) = (0,0)^T, \quad \boldsymbol{x} \in \tau,$$

where m_τ is the number of interpolation nodes belonging to the element τ.

By this point, we have obtained all expressions to calculate elements of the local stiffness matrix. Now, let us consider methods for calculation of the integral (5.1) mentioned above. Since the basis φ_{i_α} are linear functions, then $\nabla\varphi_{i_\alpha}$ are constants and (5.1) can be rewritten as

$$\bar{A}_{\alpha\beta} = c(\boldsymbol{p}_\tau)\nabla\varphi_{i_\beta} \cdot \nabla\varphi_{i_\alpha} \int_\tau d\boldsymbol{x} + b(\boldsymbol{p}_\tau) \cdot \nabla\varphi_{i_\beta} \int_\tau \varphi_{i_\alpha}d\boldsymbol{x}$$

$$+ a(\boldsymbol{p}_\tau) \int_\tau \varphi_{i_\beta}\varphi_{i_\alpha}d\boldsymbol{x} = |\tau|c(\boldsymbol{p}_\tau)\nabla\varphi_{i_\beta} \cdot \nabla\varphi_{i_\alpha}$$

$$+ b(\boldsymbol{p}_\tau) \cdot \nabla\varphi_{i_\beta} J_\tau \int_{\hat{\tau}} \hat{\varphi}_{i_\alpha}d\hat{\boldsymbol{x}} + a(\boldsymbol{p}_\tau) J_\tau \int_{\hat{\tau}} \hat{\varphi}_{i_\beta}\hat{\varphi}_{i_\alpha}d\hat{\boldsymbol{x}}. \quad (5.7)$$

First integral on the right hand side of the expression (5.7) being calculated either analytically or numerically using quadrature formula with one node results in the value of $1/6$, and we come to the expression

$$\bar{A}_{\alpha\beta} = |\tau| \left(c(\boldsymbol{p}_\tau) \nabla\varphi_{i_\beta} \cdot \nabla\varphi_{i_\alpha} + \frac{1}{3}\boldsymbol{b}(\boldsymbol{p}_\tau) \cdot \nabla\varphi_{i_\beta} \right) + a(\boldsymbol{p}_\tau) J_\tau \int_{\hat{\tau}} \hat{\varphi}_{i_\beta} \hat{\varphi}_{i_\alpha} d\hat{\boldsymbol{x}}.$$

The last integral results in different answers depending on the applied calculation methods. When using quadrature formula the answer is

$$\int_{\hat{\tau}} \hat{\varphi}_{i_\beta} \hat{\varphi}_{i_\alpha} d\hat{\boldsymbol{x}} = \hat{\varphi}_{i_\beta} \left(\hat{\boldsymbol{p}}_\tau \right) \hat{\varphi}_{i_\alpha} \left(\hat{\boldsymbol{p}}_\tau \right) \int_{\hat{\tau}} d\hat{\boldsymbol{x}} = \begin{pmatrix} \frac{1}{18} & \frac{1}{18} \\ \frac{1}{18} & \frac{1}{18} \end{pmatrix},$$

but applying analytical integration gives

$$\int_{\hat{\tau}} \hat{\varphi}_{i_\beta} \hat{\varphi}_{i_\alpha} d\hat{\boldsymbol{x}} = \begin{pmatrix} \frac{1}{12} & \frac{1}{24} \\ \frac{1}{24} & \frac{1}{12} \end{pmatrix}.$$

Next, we consider calculation of the local forcing vector on the element τ

$$\bar{F}_\alpha = \int_\tau f(\boldsymbol{x})\varphi_{i_\alpha}(\boldsymbol{x})d\boldsymbol{x}.$$

Both, analytical integration and quadrature method, lead to the same result, namely

$$\bar{F}_\alpha = \frac{1}{6} J_\tau f(\boldsymbol{p}_\tau), \quad \alpha = 1, 2, 3.$$

Regarding calculation boundary integrals

$$\bar{S}_{\alpha\beta} = \int_e \sigma(\boldsymbol{x})\varphi_{i_\alpha}\varphi_{i_\beta} d\boldsymbol{x}, \quad \bar{M}_\alpha = \int_e \mu(\boldsymbol{x})\varphi_{i_\alpha}(\boldsymbol{x})d\boldsymbol{x}, \quad (5.8)$$

which are associated with the boundary conditions on the edge e having the vertex indices i_1 and i_2, we can apply both approaches mentioned above. For the components of the vector $\boldsymbol{M}$ we come to the same expression regardless of the integration method, $\bar{M}_\alpha = |e|/2\mu(\boldsymbol{p}_e)$, $\alpha = 1, 2$, where $|e|$ is the length of the edge e and $\boldsymbol{p}_e$ is the midpoint of the edge. For the matrix $\bar{S}$ we obtain two results depending on the calculation method. In case of analytical integration we have

$$\bar{S}_{\alpha\beta} = |e|\mu(\boldsymbol{p}_e) \begin{pmatrix} \frac{1}{3} & \frac{1}{6} \\ \frac{1}{6} & \frac{1}{3} \end{pmatrix}.$$

Integration using quadrature formula gives $\bar{S}_{\alpha\beta} = (|e|/4)\mu(\boldsymbol{p}_e)$, where the fact $\varphi_{i_\alpha}(\boldsymbol{p}_e) = 1/2$ is taken into account.

5.4 CALCULATING EQUATION COEFFICIENTS

Stiffness matrices depend on the coefficients of the partial differential equation $(c, \boldsymbol{b}, a, f)$ and we need to decide how to incorporate calculation of these coefficients into FEM. Note that the equation coefficients could be piecewise functions on subdomains. Recall that the domain Ω where the solution is sought could consist of subdomains. Each subdomain is supposed to have unique label, `sdl`, and this label can be retrieved from the connectivity matrix `t` so that `sdl=t(4,:)`. As such, each element 'knows' about subdomain to which it belongs. We can construct a function which takes as an input subdomain label `sdl` and performs calculation of the coefficient, for example, if we have three subdomains Ω_1, Ω_2, Ω_3, and coefficient c is defined as

$$
c(\boldsymbol{p}) = \begin{cases} 1, & \boldsymbol{p} \in \Omega_1, \\ x_1^2 + x_2, & \boldsymbol{p} \in \Omega_2, \\ \sin(x_1 + x_2), & \boldsymbol{p} \in \Omega_3, \end{cases}
$$

then the function in listing 5.11 calculates the coefficient c.

```
1 function f=c_coeff_def(x1,x2,sdl)
2   f=zeros(size(x1)) ;
3   % 1-st subdomain
4   I=find(sdl==1); f(I)=1;
5   % 2-nd subdomain
6   I=find(sdl==2); f(I)=x1(I).^2+x2(I) ;
7   % 3-d subdomain
8   I=find(sdl==3); f(I)=sin(x1(I)+x2(I)) ;
9 end
```

Listing 5.11: MATLAB code for calculating equation coefficient.

This function can be called at any time when coefficient values need to be calculated. It is worth noting that this is a vectorized function which allows to calculate coefficients over all subdomain points.

5.5 CALCULATING GLOBAL STIFFNESS MATRIX AND FORCING VECTOR

In section 5.3 we obtained expression for calculating local stiffness matrices and local force vectors for the $\mathcal{G}_1$ mesh. Now, let us incorporate this information into the algorithms provided in sections 5.1 and 5.2. We provide two versions of the algorithm: vectorized and non-vectorized. Let us start with non-vectorized version given in listing 5.12.

```
1   function [A,F]=assemblingAF_nv(p,t,c,a,b1,b2,f)
2   % Non-vectorized assembling algorithm.
3   % The following call is allowed :
4   % A=assemblingAF_nv(p,t,c,a,b1,b2)
5   np=size(p,2) ; nt=size(t,2) ;
6   Al=zeros(nt,3,3) ; % all local stiffness matrix
7   F=zeros(np,1) ; % global force vector
8   for k=1:nt
9       % mesh point indices
10      k1=t(1,k) ; k2=t(2,k) ; k3=t(3,k) ;
11      sdl=t(4,k) ; % subdomain labels
12      % barycenter of the triangle
13      x1=(p(1,k1)+p(1,k2)+p(1,k3))/3 ;
14      x2=(p(2,k1)+p(2,k2)+p(2,k3))/3 ;
15      % gradient of the basis functions, multiplied by J
16      g1_x1=p(2,k2)—p(2,k3) ; g1_x2=p(1,k3)—p(1,k2) ;
17      g2_x1=p(2,k3)—p(2,k1) ; g2_x2=p(1,k1)—p(1,k3) ;
18      g3_x1=p(2,k1)—p(2,k2) ; g3_x2=p(1,k2)—p(1,k1) ;
19      J=abs(g3_x2.*g2_x1—g3_x1.*g2_x2) ; % J=2*area
20      % evaluate c , b , a on triangles barycenter
21      cf=feval(c,x1,x2,sdl) ;
22      af=feval(a,x1,x2,sdl) ;
23      b1f=feval(b1,x1,x2,sdl) ;
24      b2f=feval(b2,x1,x2,sdl) ;
25      % diagonal and off-diagonal elements of mass matrix
26      ao=(af/24).*J ; ad=4*ao ; % 'exact' integration
27      % ao=(af/18).*J ; ad=3*ao ; % quadrature rule
28      % b contributions
29      b1f=b1f/6 ; b2f=b2f/6 ;
30      bg1=b1f.*g1_x1+b2f.*g1_x2 ;
31      bg2=b1f.*g2_x1+b2f.*g2_x2 ;
32      bg3=b1f.*g3_x1+b2f.*g3_x2 ;
33      % coefficients of the stiffness matrix
34      cf=(0.5*cf)./J ;
35      a12=cf.*(g1_x1.*g2_x1+g1_x2.*g2_x2)+ao ;
36      a23=cf.*(g2_x1.*g3_x1+g2_x2.*g3_x2)+ao ;
37      a31=cf.*(g3_x1.*g1_x1+g3_x2.*g1_x2)+ao ;
```

```
38     Al(k,:,:)=[ad—a31—a12+bg1 a12+bg2 a31+bg3
39                a12+bg1 ad—a12—a23+bg2 a23+bg3
40                a31+bg1 a23+bg2 ad—a23—a31+bg3] ;
41     if nargout==2
42       ff=feval(f,x1,x2,sdl) ;
43       I=[k1;k2;k3] ;
44       F(I)=F(I)+ff*J/6 ;
45     end
46   end
47   A=sparse(np,np) ; % global stiffness matrix
48   for i=1:3
49     for j=1:3
50       A=A+sparse(t(i,:),t(j,:),Al(:,i,j),np,np) ;
51     end
52   end
53   if nargout==1
54     F=[] ;
55   end
56 end
```

Listing 5.12: MATLAB code for non-vectorized version of the algorithm for calculating elements and assembling FEM matrices.

Analysis of the function assemblingAF_nv() shows that in order to vectorize it we need to avoid using loop and assemble local stiffness matrices on place once they have been calculated. As a result, we come to the vectorized version of the algorithm (see listing 5.13).

```
1  function [A,F]=assemblingAF(p,t,c,a,b1,b2,f)
2  % Vectorized assembling algorithm.
3  % The following call is allowed :
4  % A=assemblingAF(p,t,c,a,b1,b2)
5  np=size(p,2) ;
6  k1=t(1,:) ; k2=t(2,:) ; k3=t(3,:) ; % mesh point indices
7  sdl=t(4,:) ; % subdomain labels
8  % barycenter of the triangles
9  x1=(p(1,k1)+p(1,k2)+p(1,k3))/3 ;
10 x2=(p(2,k1)+p(2,k2)+p(2,k3))/3 ;
11 % gradient of the basis functions, multiplied by J
12 g1_x1=p(2,k2)—p(2,k3) ; g1_x2=p(1,k3)—p(1,k2) ;
13 g2_x1=p(2,k3)—p(2,k1) ; g2_x2=p(1,k1)—p(1,k3) ;
14 g3_x1=p(2,k1)—p(2,k2) ; g3_x2=p(1,k2)—p(1,k1) ;
```

```
15    J=abs(g3_x2.*g2_x1—g3_x1.*g2_x2) ; % J=2*area
16    % evaluate c , b , a on triangles barycenter
17    cf=feval(c,x1,x2,sdl) ;
18    af=feval(a,x1,x2,sdl) ;
19    b1f=feval(b1,x1,x2,sdl) ;
20    b2f=feval(b2,x1,x2,sdl) ;
21    % diagonal and off diagonal elements of mass matrix
22    ao=(af/24).*J ; ad=4*ao ; % 'exact' integration
23    % ao=(af/18).*J ; ad=3*ao ; % quadrature rule
24    % coefficients of the stiffness matrix
25    cf =(0.5*cf)./J ;
26    a12=cf.*(g1_x1.*g2_x1+g1_x2.*g2_x2)+ao ;
27    a23=cf.*(g2_x1.*g3_x1+g2_x2.*g3_x2)+ao ;
28    a31=cf.*(g3_x1.*g1_x1+g3_x2.*g1_x2)+ao ;
29    if all(b1f==0) && all( b2f==0) % symmetric problem
30      A=sparse(k1,k2,a12,np,np) ;
31      A=A+sparse(k2,k3,a23,np,np) ;
32      A=A+sparse(k3,k1,a31,np,np) ;
33      A=A+A.';
34      A=A+sparse(k1,k1,ad—a31—a12,np,np) ;
35      A=A+sparse(k2,k2,ad—a12—a23,np,np) ;
36      A=A+sparse(k3,k3,ad—a23—a31,np,np) ;
37    else
38      % b contributions
39      b1f=b1f/6 ; b2f=b2f/6 ;
40      bg1=b1f.*g1_x1+b2f.*g1_x2 ;
41      bg2=b1f.*g2_x1+b2f.*g2_x2 ;
42      bg3=b1f.*g3_x1+b2f.*g3_x2 ;
43      A=sparse(k1,k2,a12+bg2,np,np) ;
44      A=A+sparse(k2,k3,a23+bg3,np,np) ;
45      A=A+sparse(k3,k1,a31+bg1,np,np) ;
46      A=A+sparse(k2,k1,a12+bg1,np,np) ;
47      A=A+sparse(k3,k2,a23+bg2,np,np) ;
48      A=A+sparse(k1,k3,a31+bg3,np,np) ;
49      A=A+sparse(k1,k1,ad—a31—a12+bg1,np,np) ;
50      A=A+sparse(k2,k2,ad—a12—a23+bg2,np,np) ;
51      A=A+sparse(k3,k3,ad—a23—a31+bg3,np,np) ;
52    end
53    if nargout==2
54      ff=feval(f,x1,x2,sdl) ;
```

```
55      ff=(ff/6).*J ;
56      F=sparse(k1,1,ff,np,1) ;
57      F=F+sparse(k2,1,ff,np,1) ;
58      F=F+sparse(k3,1,ff,np,1) ;
59    else
60      F=[] ;
61    end
62 end
```

Listing 5.13: MATLAB code for the vectorized version of the algorithm for calculating elements and assembling FEM matrices.

Let us compare computational costs of these two functions. We take a domain having three subdomains with geometry defined by matrix geometryMatrix (see figure 4.1) and specify coefficients of the PDE as follows:

```
1 function f=a_coeff(x1,x2,sdl)
2    f=zeros(size(x1)) ;
3    % 1-st subdomain
4    I=find(sdl==1); f(I)=x2(I).^2;
5    % 2-nd subdomain
6    I=find(sdl==2); f(I)=x2(I).^2 ;
7    % 3-d subdomain
8    I=find(sdl==3); f(I)=x2(I).^2 ;
9 end
```

Listing 5.14: Coding coefficient a of the PDE.

```
1 function f=b1_coeff(x1,x2,sdl)
2    f=zeros(size(x1)) ;
3    % 1-st subdomain
4    I=find(sdl==1); f(I)=x1(I)+x2(I);
5    % 2-nd subdomain
6    I=find(sdl==2); f(I)=x1(I)—x2(I) ;
7    % 3-d subdomain
8    I=find(sdl==3); f(I)=x1(I).*x2(I) ;
9 end
```

Listing 5.15: Coding first component of the coefficient b of the PDE.

```
1 function f=b2_coeff(x1,x2,sdl)
2    f=zeros(size(x1)) ;
```

```
3    % 1-st subdomain
4    I=find(sdl==1); f(I)=1;
5    % 2-nd subdomain
6    I=find(sdl==2); f(I)=1 ;
7    % 3-d subdomain
8    I=find(sdl==3); f(I)=1 ;
9 end
```

Listing 5.16: Coding second component of the coefficient b of the PDE.

```
1 function f=c_coeff(x1,x2,sdl)
2    f=zeros(size(x1)) ;
3    % 1-st subdomain
4    I=find(sdl==1); f(I)=1;
5    % 2-nd subdomain
6    I=find(sdl==2); f(I)=x1(I).^2+x2(I) ;
7    % 3-d subdomain
8    I=find(sdl==3); f(I)=sin(x1(I)+x2(I)) ;
9 end
```

Listing 5.17: Coding coefficient c of the PDE.

```
1 function f=f_coeff(x1,x2,sdl)
2    f=zeros(size(x1)) ;
3    % 1-st subdomain
4    I=find(sdl==1); f(I)=1;
5    % 2-nd subdomain
6    I=find(sdl==2); f(I)=x1(I).^2+x2(I) ;
7    % 3-d subdomain
8    I=find(sdl==3); f(I)=sin(x1(I)+x2(I)) ;
9 end
```

Listing 5.18: Coding right hand side, f, of the PDE.

Then, we run the script shown in listing 5.19 to measure CPU time for three meshes with different number of nodes.

```
1 function testAssemblingAF
2    time=[] ; time_nv=[] ; % assembling time
3    nt=[] ; np=[] ;
4    g=geometryMatrix;
5    % PDE coefficients;
6    c=@c_coeff ;
```

```
7   a=@a_coeff;
8   b1=@b1_coeff ;
9   b2=@b2_coeff;
10  f=@f_coeff ;
11  for h=[0.1 0.05 0.02]
12      [p,e,t]=initmesh(g,'hmax',h) ;
13      np=[np size(p,2)] ; nt=[nt size(t,2)] ;
14      tic ;
15      [A,F]=assemblingAF(p,t,c,a,b1,b2,f) ;
16      time=[time toc];
17      tic ;
18      [A,F]=assemblingAF_nv(p,t,c,a,b1,b2,f) ;
19      time_nv=[time_nv toc];
20  end
21  disp (' np nt assemblingAF assemblingAF_nv')
22  disp([np' nt' time' time_nv'])
23 end
```

Listing 5.19: MATLAB code for testing efficiency of vectorized and non-vectorized FEM assembling algorithms.

As a result, we come to table 5.3 which shows that vectorized version outperforms non-vectorized one, that is predictable due to more efficient memory management of the vectorized calculations in MATLAB.

TABLE 5.3: Summary of the computational efficiency test for assembling global stiffness matrix and forcing vector. Last two columns contain CPU time for the corresponding algorithm version.

np	nt	assemblingAF	assemblingAF_nv
704	1312	0.00391	1.08008
2670	5148	0.00781	3.73047
16473	32466	0.05566	23.62988

Python implementation of the function assemblingAF() can be found in Appendix E.1 where corresponding explanations are provided.

5.6 CALCULATING BOUNDARY CONDITIONS

Below, we describe one of the possible methods of calculating boundary conditions. Later, we make use of this approach to incorporate boundary conditions into final FEM matrix.

Recall that the boundary Γ of the problem domain Ω is supposed to be split into two parts, Γ_D and Γ_R, on which different boundary conditions could be defined (see section 3.1). Boundary parts Γ_D and Γ_R, in turn, are union of the corresponding boundary segments defined in geometry matrix. We associate each segment with either function u_D (for the Dirichlet boundary Γ_D) or with pair of functions (σ, μ) (for the Robin boundary Γ_R). Thereby, we define conditions on Γ_D and Γ_R separately.

Suppose that Γ_D is non-empty and consists of the boundary segments with indices $(i_1, i_2, \ldots, i_d)$ and function u_D could be different for each segment. We define a function uD(x1,x2,sdl) in the way described in section 5.4. As a result, Dirichlet boundary conditions are defined by the list of boundary segment indices bsD=[i1 , i2 , .. , id] and by corresponding function (likewise in section 5.4). Note that we need to bear in mind that sdl parameter in a function has now meaning of the local index. For instance, supposing we have three Dirichlet boundary segments with indices 5, 2 and 7, then bsD = [5 2 7]. If $u_D = 1$ for the fifth and seventh segments, $u_D = x_1 + x_2$ for the second segment, then we can define the function uD as shown in listing 5.20.

```
1 function f=uD(x1,x2,sdl)
2    f=zeros(size(x1)) ;
3    % 1-st segment (local enumeration)
4    I=find(sdl==1); f(I)=1;
5    % 2-nd segment
6    I=find(sdl==2); f(I)=x1(I)+x2(I) ;
7    % 3-d segment
8    I=find(sdl==3); f(I)=1;
9 end
```

Listing 5.20: MATLAB code for defining Dirichlet BC.

If there is no Dirichlet boundary conditions, then bcD=[] and uD=[]; if $\Gamma_D = \Gamma$ then bsD=inf.

Likewise, we define boundary conditions on Γ_R through the list of indices bsR and corresponding functions sg and mu for (σ, μ). In this case, if $\sigma = \mu = 0$ then this segment is not indicated in the list; if $\Gamma_R = \Gamma$ then bsR=inf. Finally, if $\Gamma_D = \emptyset$, $\Gamma_R = \Gamma$ and $\sigma = \mu = 0$ then bsN=inf, sg=[], mu=[].

For convenience, we pack data (bsD,uD,bsR,sg,mu) into structure bc with corresponding fields.

5.7   ASSEMBLING BOUNDARY CONDITIONS

Now, we gather all the necessary data to form the FEM system defined by (3.7)

$$\sum_{j\in\mathcal{I}_U} \tilde{A}_{ij}u_j = q_i, \quad q_i = \tilde{F}_i - \sum_{j\in\mathcal{I}_D} \tilde{A}_{ij}u_{D_j}, \ i\in\mathcal{I}_U,$$

where the matrix $\tilde{A} = \{\tilde{A}_{ij}\}_{i,j=1}^{n_p} = A + S$, $\tilde{F} = \{\tilde{F}_i\}_{i=1}^{n_p} = F + M$, $\mathcal{I}_D$ are the indices of the points with Dirichlet boundary conditions, and $\mathcal{I}_U$ are the rest points.

Function `assemblingAF()` (see section 5.5) calculates matrix A and vector F and we need a similar function to calculate S and M. In addition, this function would enable us calculating vector U_D with n_p elements such that all its elements are equal to zero apart from elements with indices $j \in \mathcal{I}_D$ having value of u_{D_j}.

Matrix $\tilde{A}_{\mathcal{I}_U} = \{\tilde{A}\}_{i,j\in\mathcal{I}_U}$ is obtained from the matrix $\tilde{A}$ by deleting rows and columns with indices belonging to the set $\mathcal{I}_D$. To perform this operation we introduce a matrix $N \sim (n_p \times m)$ where m is the number of elements in the set $\mathcal{I}_U$ and

$$N^T \cdot (u_1, u_2, \cdots, u_{n_p})^T = (u_{i_1}, u_{i_2}, \cdots, u_{i_U})^T.$$

Matrix N can be defined as `N=sparse([i1,i2,..,in],1:m,1,np,m)`. Consequently, if S, M, N, and u_D are calculated, then matrix $\tilde{A}$ and vector q are defined through the following expressions

$$\tilde{A}_{\mathcal{I}_U} = N^T(A+S)N, \quad q = N^T((F+M) - (A+S)U_D).$$

Note that after solving the system $\tilde{A}_{\mathcal{I}_U} u = q$ the final vector of solution at the node points can be obtained as $U = Nu + U_D$ (complementing vector u by the boundary values U_D). As a result, we come to the function `assemblingBC()` in listing 5.21 which calculates contributions from the boundary elements.

```
1  function [N,S,UD,M]=assemblingBC(bc,p,e)
2    % The following call is also allowed :
3    % [N,S]=assemblingBC(bc,p,e)
4    % [N,S,UD]=assemblingBC(bc,p,e)
5    np=size(p,2) ;
6    S=sparse(np,np) ; M=sparse(np,1) ; N=speye(np,np) ;
7    UD=sparse(np,1) ;
```

```
8    if all(bc.bsR==inf) && (numel(bc.sg)==0) && (numel(bc.mu)==0)
9        return % all boundary are homogeneous Robin BC
10   end
11   e=e(:,e(6,:)==0 | e (7,:)==0) ; % all boundary edges
12   k=e(5,:) ;
13   [bsg,~]=find(sparse(k,k,1,np,np)) ; % boundary segment indices
14   nbsg=numel(bsg) ;
15   % set localal enumeration of boundary segments
16   local=zeros(1,nbsg) ;
17   % set Robin boundary conditions
18   if nargout>=2
19   bsR=bc.bsR ;
20   if numel(bsR)>0
21     if bsR==inf , bsR=bsg ; end % all BC are mixed
22       local(bsR)=1:numel(bsR) ;
23       % find boudary edges with Robin b.c.
24       eR=e([1 2 5],ismember(k,bsR)) ;
25       k1=eR(1,:) ; k2=eR(2,:) ; % corner point indices
26       x1=0.5*(p(1,k1)+p(1,k2)) ; x2=0.5*(p(2,k1)+p(2,k2)) ;
27       h=sqrt((p(1,k2)—p(1,k1)).^2+(p(2,k2)—p(2,k1)).^2) ;
28       sdl=local(eR(3,:)) ;
29       % evaluate sigma and mu on edges barycenter
30       sf=feval(bc.sg,x1,x2,sdl) ;
31       mf=feval(bc.mu,x1,x2,sdl) ;
32       % diagonal and off diagonal elements
33       so=(sf/6).*h ; sd=2*so ; % 'exact' integration
34       %so=(sf/4).*h ; sd = so ; % quadrature rule
35       S=sparse(k1,k2,so,np,np) ;
36       S=S+sparse(k2,k1,so,np,np) ;
37       S=S+sparse(k1,k1,sd,np,np) ;
38       S=S+sparse(k2,k2,sd,np,np) ;
39       if nargout==4
40         mf=feval(bc.mu,x1,x2,sdl) ;
41         mf=(mf/2).*h ;
42         M=sparse(k1,1,mf,np,1) ;
43         M=M+sparse(k2,1,mf,np,1) ;
44       end
45     end
46   end
47   % set Dirichlet boundary conditions
```

```
48   bsD=bc.bsD ;
49   if numel (bsD)>0
50     if bsD==inf
51       bsD=bsg ;
52     end % all b.c. are the Dirichlet one
53     local(bsD)=1:numel(bsD) ;
54     if all(local==0)
55       disp ('error. bsD+bsR~=number of boundary segments')
56     end
57     eD=e([1 2 5],ismember(k,bsD)) ; % Dirichlet BC edges
58     sdl=local(eD(3,:)) ;
59     k1=eD(1,:) ; k2=eD(2,:) ; % indices of corner points
60     iD=[k1 k2] ;
61     [id,~]=find(sparse(iD,iD,1,np,np)) ; % Dirichlet point indices
62     iN=ones(1,np) ; iN(id)=zeros(1,numel(id)) ;
63     iN=find(iN) ; % indices of non-Dirichlet points
64     niN=numel(iN) ;
65     N=sparse(iN,1:niN,1,np,niN) ;
66     if nargout>=3 % evaluate UD on Dirichlet points
67       UD(k1)=feval(bc.uD,p(1,k1),p(2,k1),sdl) ;
68       UD(k2)=feval(bc.uD,p(1,k2),p(2,k2),sdl) ;
69     end
70   end
71 end
```

Listing 5.21: MATLAB code for calculating and assembling BC.

For clarification, MATLAB function t=ismember(a,b) returns Boolean vector such that t(k)=1 if a(k) ∈ b and t(k)=0 if a(k) ∉ b.

Functions described above allow us forming and solving FEM system. As a final step, we provide a function assemblingPDE() in listing 5.22 which assembles a PDE problem and solves it using FEM. Depending on the number of output arguments, it returns either FEM matrices or a final solution vector.

```
1 function [A,F,N,UD,S,M]=assemblingPDE(bc,p,e,t,c,a,b1,b2,f)
2 %[A,F,N,UD,S,M]=assemblingPDE(bc,p,e,t,c,a,b1,b2,f)  - returns
3 % FEM matrices.
4 % u=assemblingPDE(bc,p,e,t,c,a,b1,b2,f) - returns
5 % FEM solution represented as a column vector.
6 if nargout==6
7   [A,F]=assemblingAF(p,t,c,a,b1,b2,f) ;
```

```
8       [N,S,UD,M]=assemblingBC(bc,p,e) ;
9   elseif nargout==1
10      [A,F]=assemblingAF(p,t,c,a,b1,b2,f) ;
11      [N,S,UD,M]=assemblingBC(bc,p,e) ;
12      if size(N,2)==size(p,2) % no Dirichlet boundary conditions
13         A=A+S;
14         F=F+M;
15      else
16         Nt=N.';
17         A=A+S;
18         F=Nt*((F+M)—A*UD) ;
19         A=Nt*A*N;
20      end
21      u=A\F;
22      A=N*u+UD; % returns A=u
23   else
24      error('Wrong number of output parameters.' ) ;
25   end
26 end
```

Listing 5.22: MATLAB code for assembling FEM matrices and solving a PDE problem.

Functions `assemblingAF()` and `assemblingPDE()` translated into Python programming language can be found in Appendix E.2.

5.8 SOLVING EXAMPLE PROBLEM

Let us apply function `assemblingPDE()` to solve an example problem. We consider PDE of the form

$$-\boldsymbol{\nabla} \cdot (c(\boldsymbol{x})\nabla u) + \boldsymbol{b}(\boldsymbol{x}) \cdot \nabla u + a(\boldsymbol{x})u = f(\boldsymbol{x}), \quad \boldsymbol{x} = (x_1, x_2) \in \Omega,$$

with boundary conditions

$$u(\boldsymbol{x}) = u_D(\boldsymbol{x}), \ \boldsymbol{x} \in \Gamma_D;$$
$$c(\boldsymbol{x})\nabla u \cdot \boldsymbol{\nu}(\boldsymbol{x}) + \sigma(\boldsymbol{x})u = \mu(\boldsymbol{x}), \ \boldsymbol{x} \in \Gamma_R,$$

where Ω is a circle of the unity radius. Figure 5.3 shows boundaries of the circle having four segments, where Γ_D comprises segments 1 and 3, and Γ_R comprises segments 2 and 4.

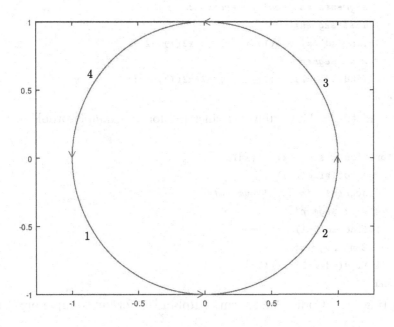

FIGURE 5.3: Geometry for the example problem.

We define coefficients

$$c = 2 + x_1 + x_2,$$
$$a = x_1 + x_2,$$
$$\boldsymbol{b} = (x_1, x_2),$$
$$f = -8 - 6(x_1 + x_2) + (2 + x_1 + x_2)(x_1^2 + x_2^2),$$

and boundary conditions

$$u_D = x_1^2 + x_2^2,$$

$$\sigma = \begin{cases} 0, & \text{on } 2^{nd} \text{ segment,} \\ 2, & \text{on } 4^{th} \text{ segment ,} \end{cases}$$

$$\mu = \begin{cases} 2(2 + x_1 + x_2)u_D, & \text{on } 2^{nd} \text{ segment,} \\ 2(2 + x_1 + x_2 + 1)u_D, & \text{on } 4^{th} \text{ segment.} \end{cases}$$

Boundary conditions σ and μ can be coded through the functions

```
1 function f=u_D(x1,x2,sdl)
2    f=zeros(size(x1)) ;
```

```
3    % segments in local numeration:
4    % 1-st segment
5    I=find(sdl==1); f(I)=x1(I).^2+x2(I).^2;
6    % 2-nd segment
7    I=find(sdl==2); f(I)=x1(I).^2+x2(I).^2 ;
8  end
```

Listing 5.23: Coding Dirichlet BC for an example problem.

```
1  function f=sigma(x1,x2,sdl)
2    f=zeros(size(x1)) ;
3    % segments in local numeration:
4    % 1-st segment
5    I=find(sdl==1); f(I)=0;
6    % 2-nd segment
7    I=find(sdl==2); f(I)=2 ;
8  end
```

Listing 5.24: Coding first term of Robin BC for an example problem.

```
1  function f=mu(x1,x2,sdl)
2    f=zeros(size(x1)) ;
3    % segments in local numeration:
4    % 1-st segment
5    I=find(sdl==1);
6    f(I)=2*(2+x1(I)+x2(I)).*(x1(I).^2+x2(I).^2) ;
7    % 2-nd segment
8    I=find(sdl==2);
9    f(I)=2*(2+x1(I)+x2(I)+1).*(x1(I).^2+x2(I).^2) ;
10 end
```

Listing 5.25: Coding second term of Robin BC for an example problem.

It is easy to see that solution to this problem is given by function $u = x_1^2 + x_2^2$. We utilize function assemblingPDE() to solve this problem numerically on the meshes with h = [0.5 0.1 0.05 0.02]. Since the exact solution is known we can find an absolute error

$$err = \max_{x \in \mathcal{T}_h} |u(x) - u_h(x)|,$$

and relative error, $err_h = err/h^2$. Function testPDE in listing 5.26 solves this problem and calculates FEM solution errors.

```
 1 function testPDE
 2   clear all ;close all ; clc
 3   g='circleg' ; np=[] ; nt=[] ; errh2=[] ; err=[] ;
 4   t_assemblingAF=[] ; t_assemblingBC=[] ; t_solving=[] ;
 5   for h=[0.5 0.1 0.05 0.02]
 6     [p,e,t]=initmesh(g,'hmax',h) ;
 7     exact=@(x1,x2,sdl) x1.^2+x2.^2; % exact solution
 8     x1=p(1,:) ; x2=p(2,:) ;
 9     u_exact=exact(x1,x2,1)';
10     % PDE coefficients
11     c=@(x1,x2,sdl) 2+x1+x2 ;
12     a=@(x1,x2,sdl) x1+x2 ;
13     f=@(x1,x2,sdl) -8-6*(x1+x2)+(2+x1+x2).*(x1.^2+x2.^2) ;
14     b1=@(x1,x2,sdl) x1 ;
15     b2=@(x1,x2,sdl) x2 ;
16     % boundary conditions
17     bc.bsD=[1 3] ; bc.uD=@u_D ;
18     bc.bsR=[2 4] ; bc.sg=@sigma ; bc.mu=@mu ;
19     tic
20     [A,F]=assemblingAF(p,t,c,a,b1,b2,f) ;
21     t_assemblingAF=[t_assemblingAF toc] ;
22     tic
23     [N,S,UD,M]=assemblingBC(bc,p,e) ;
24     t_assemblingBC=[t_assemblingBC toc] ;
25     tic
26     u=assemblingPDE(bc,p,e,t,c,a,b1,b2,f) ;
27     t_solving =[t_solving toc] ;
28     np=[np size(p,2)] ; nt=[nt size(t,2)] ;
29     err=[err norm(u_exact-u,inf)] ;
30     errh2=[errh2 norm(u_exact-u,inf)/h^2] ;
31   end
32   disp ('np    assemblingAF    assemblingBC    solving')
33   disp ( [np' t_assemblingAF' t_assemblingBC' t_solving'] )
34   err
35   errh2
36 end
```

Listing 5.26: MATLAB code for solving example problem.

As a result we obtain tables 5.4 and 5.5 summarizing testing data.
Data for error magnitudes in table 5.5 indicate that the problem has

TABLE 5.4: Summary of the solving test problem. Last three columns contain CPU time for the corresponding functions.

np	assemblingAF	assemblingBC	solving
33	0.00195	0.00586	0.00293
544	0.00195	0.00195	0.00683
2169	0.00586	0.00195	0.02344
13525	0.05859	0.00683	0.31348

TABLE 5.5: Errors when solving the test problem numerically.

np	33	544	2169	13525
err	0.07266	0.00252	0.00093	0.00013
errh2	0.29066	0.25173	0.37120	0.32498

been solved successfully. Absolute error decreases with mesh resolution refining while relative error remains stable with slight fluctuation.

Python version of the code for solving example problem can be found in Appendix E.3.

CHAPTER 6

Nonlinear Basis Functions

H ERE we consider $\mathcal{G}_2$ mesh and obtain expressions for calculation of basis functions for mesh triangular elements with straight as well as curvilinear edges. Analysis of how to deduce expression for calculating elements of stiffness matrix is given here. We provide mathematical formulations and expressions without giving MATLAB implementations codes. Based on the content of this chapter, coding functions can be easily composed. Basically, we outline techniques to extend existing FEM framework for $\mathcal{G}_1$ mesh to FEM with $\mathcal{G}_2$ mesh.

6.1 QUADRATIC BASIS FUNCTIONS FOR LINEAR TRIANGULAR ELEMENTS

Methodology to obtain basis functions for the $\mathcal{G}_2$ triangles with straight edges is similar to one of $\mathcal{G}_1$ meshes with the difference in number of interpolation nodes.

Let us consider a canonical triangle element $\hat{\tau}$ in the coordinate system $\hat{x} = (\hat{x}_1, \hat{x}_2)$ having six interpolation nodes $\hat{p}_i$, $i = 1, \ldots, 6$ (here $\hat{p}_4, \hat{p}_5$, and $\hat{p}_6$ are midpoints). Then $\mathcal{G}_2$ are all polynomials of the second order on the element $\hat{\tau}$ (see figure 6.1 with original triangle element and corresponding basis element).

If $\hat{x}_3 = 1 - \hat{x}_1 - \hat{x}_2$, then basis in $\hat{\mathcal{G}}_2$ is defined by the following functions

$$\hat{\varphi}_1 = \hat{x}_3(2\hat{x}_3 - 1), \quad \hat{\varphi}_2 = \hat{x}_1(2\hat{x}_1 - 1), \quad \hat{\varphi}_3 = \hat{x}_2\left(2\hat{x}_2 - \frac{1}{2}\right),$$

$$\hat{\varphi}_4 = 4\hat{x}_1\hat{x}_3, \quad \hat{\varphi}_5 = 4\hat{x}_1\hat{x}_2, \quad \hat{\varphi}_6 = 4\hat{x}_2\hat{x}_3, \tag{6.1}$$

DOI: 10.1201/9781003265979-6

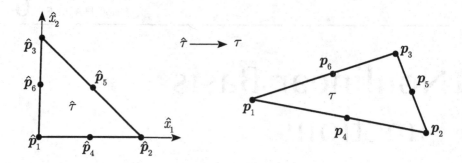

FIGURE 6.1: Canonical element $\hat{\tau}$ and original $\mathcal{G}_2$ mesh element τ.

and for arbitrary function $\hat{z} \in \hat{\mathcal{G}}_2$ is valid

$$\hat{z}(\hat{\boldsymbol{x}}) = \sum_{i=1}^{6} \hat{z}(\hat{\boldsymbol{p}}_i)\hat{\varphi}_i(\hat{\boldsymbol{x}}), \quad \hat{\boldsymbol{x}} \in \hat{\tau}.$$

Coordinate transformation $\hat{\tau} \to \tau$ is defined by (see (5.4) – (5.6))

$$\boldsymbol{x} = B_\tau \hat{\boldsymbol{x}} + \boldsymbol{p}_1, \tag{6.2}$$

and, in addition, points $\hat{\boldsymbol{p}}_i$ are transformed into points $\boldsymbol{p}_i$, $i = 1, \ldots, 6$. Inverse transformation is defined by $\hat{\boldsymbol{x}} = B_\tau^{-1}(\boldsymbol{x} - \boldsymbol{p}_1)$ and functions $\varphi_i(\boldsymbol{x}) = \hat{\varphi}_i(B_\tau^{-1}(\boldsymbol{x} - \boldsymbol{p}_1))$ represent a basis in $\mathcal{G}_2$. As a result, $\hat{z}(\hat{\boldsymbol{x}})$ transforms into $z(\boldsymbol{x})$ such that

$$z(\boldsymbol{x}) = \sum_{i=1}^{6} z(\boldsymbol{p}_i)\varphi_i(\boldsymbol{x}), \quad \boldsymbol{x} \in \tau. \tag{6.3}$$

It is worth noting here that unlike $\mathcal{G}_1$ mesh, basis functions for $\mathcal{G}_2$ mesh are nonlinear functions, namely, of the second order. This fact results in slightly more complicated calculations of its gradients.

6.2 QUADRATIC BASIS FUNCTIONS FOR CURVILINEAR TRIANGULAR ELEMENTS

Usually, curvilinear elements are used as mesh elements adjoining the boundary to approximate curvilinear geometry in most precise way. Figure 6.2 shows $\mathcal{G}_2$ canonical element $\hat{\tau}$ (basis functions for $\hat{\tau}$ are defined by (6.1)) and corresponding curvilinear element τ. Points $\boldsymbol{p}_i$, $i = 1, \ldots, 6$ form the element so that each three points define a parabolic curve.

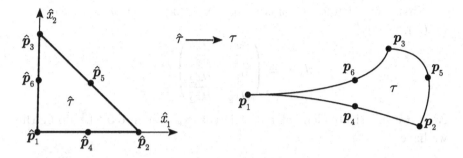

FIGURE 6.2: Canonical element $\hat{\tau}$ and original curvilinear $\mathcal{G}_2$ mesh element τ.

For example, it is possible to plot a unique parabola through nodes $\boldsymbol{p}_2$, $\boldsymbol{p}_3$ and $\boldsymbol{p}_5$. As a result, the element τ can be expressed through the functions of second order, or, in other words, through basis functions (6.1). We come to the transformation

$$\boldsymbol{x} = \sum_{i=1}^{6} \boldsymbol{p}_i \hat{\varphi}_i(\hat{\boldsymbol{x}}), \quad \hat{\boldsymbol{x}} \in \hat{\tau} \tag{6.4}$$

which transforms $\hat{\tau}$ into τ, and points $\hat{\boldsymbol{p}}_i$ are transformed into $\boldsymbol{p}_i$. Note that if nodes of the element τ lie on the straight line, then transformation (6.4) reduces to linear form (6.2). As such, the difference between linear and curvilinear triangle elements is the way of transformation from $\hat{\tau}$ to τ.

Likewise linear $\mathcal{G}_2$ elements, to represent any function defined on curvilinear element τ through the basis function we utilize the expression (6.3) where basis functions now are not polynomials of the second order but function of the $\boldsymbol{x}$ being itself second-order polynomial.

6.3 STIFFNESS MATRIX WITH QUADRATIC BASIS FUNCTIONS

Let us consider calculation of the local stiffness matrix defined on the $\mathcal{G}_2$ mesh elements (see (5.1))

$$\bar{A}_{\alpha\beta} = \int_{\tau} (c\nabla\varphi_\beta \cdot \nabla\varphi_\alpha + \boldsymbol{b} \cdot \nabla\varphi_\beta\varphi_\alpha + a\varphi_\beta\varphi_\alpha) \, d\boldsymbol{x}. \tag{6.5}$$

Method of calculation of the integral (6.5) considered below is applicable to both, curvilinear and linear, $\mathcal{G}_2$ elements.

First, we define Jacobian matrix $J(\boldsymbol{x})$ with $\boldsymbol{x} = \boldsymbol{x}(\hat{\boldsymbol{x}}) = (x_1(\hat{\boldsymbol{x}}), x_2(\hat{\boldsymbol{x}}))^T$,

$$J(\hat{\boldsymbol{x}}) = \begin{pmatrix} \dfrac{\partial x_1}{\partial \hat{x}_1} & \dfrac{\partial x_2}{\partial \hat{x}_1} \\ \dfrac{\partial x_1}{\partial \hat{x}_2} & \dfrac{\partial x_2}{\partial \hat{x}_2} \end{pmatrix}(\hat{\boldsymbol{x}}).$$

Next, we substitute element τ by $\hat{\tau}$ in the expression (6.5). By doing this we have

$$c(\boldsymbol{x}) \to \hat{c}(\hat{\boldsymbol{x}}) = c(\boldsymbol{x}(\hat{\boldsymbol{x}})), \quad \varphi_\alpha(\boldsymbol{x}) \to \hat{\varphi}_\alpha(\hat{\boldsymbol{x}})$$

$$\nabla \to J^{-1}(\hat{\boldsymbol{x}})\hat{\nabla}, \quad \hat{\nabla} = (\partial/\partial\hat{x}_1, \partial/\partial\hat{x}_2)^T, \quad d\boldsymbol{x} = |J(\hat{\boldsymbol{x}})|d\hat{\boldsymbol{x}},$$

where basis functions $\hat{\varphi}_\alpha(\hat{\boldsymbol{x}})$ are defined by (6.1), and

$$\bar{A}_{\alpha\beta} = \int_{\hat{\tau}} \Big(\hat{c}(\hat{\boldsymbol{x}}) J^{-1}(\hat{\boldsymbol{x}})\hat{\nabla}\hat{\varphi}_\beta \cdot J^{-1}(\hat{\boldsymbol{x}})\hat{\nabla}\hat{\varphi}_\alpha$$

$$+ \hat{\boldsymbol{b}}(\hat{\boldsymbol{x}}) \cdot J^{-1}(\hat{\boldsymbol{x}})\hat{\nabla}\hat{\varphi}_\beta\hat{\varphi}_\alpha + \hat{a}(\hat{\boldsymbol{x}})\hat{\varphi}_\beta\hat{\varphi}_\alpha \Big) |J(\hat{\boldsymbol{x}})|d\hat{\boldsymbol{x}}. \quad (6.6)$$

To compute integral (6.6) we apply quadrature rule. Let us denote the integrand as $\hat{z}(\hat{\boldsymbol{x}})$, then

$$\bar{A}_{\alpha\beta} = \int_{\hat{\tau}} \hat{z}(\hat{\boldsymbol{x}})d\hat{\boldsymbol{x}} \approx \sum_{i=1}^{q} \hat{\gamma}_i \hat{z}(\hat{\boldsymbol{d}}_i),$$

where coefficients $\hat{\gamma}_i$ and nodes $\hat{\boldsymbol{d}}_i$, $i = 1, \ldots, q$ depend on particular chosen quadrature rule. Using quadrature with polynomials of the second order suffice for $\mathcal{G}_2$ mesh.

When calculating integral (6.6) we make use of some properties of the $\mathcal{G}_2$ mesh. Firstly, when computing $\bar{A}_{\alpha\beta}$ over triangle with straight edges, matrix $J^{-1} = B^{-T}$ does not depend on $\hat{\boldsymbol{x}}$ and $\bar{A}_{\alpha\beta}$ can be computed only once. However, in case of curvilinear triangular elements, Jacobian $J^{-1}(\hat{\boldsymbol{x}})$ needs to be calculated q times. Secondly, we can facilitate calculation of (6.6) by introducing matrices $\hat{D}(\hat{\boldsymbol{d}}_i) \sim (2 \times 6)$

$$\hat{D}(\hat{\boldsymbol{d}}_i) = \left\{ \hat{\nabla}\hat{\varphi}_\beta(\hat{\boldsymbol{d}}_i) \right\}_{\beta=1}^{6}, \quad D(\hat{\boldsymbol{d}}_i) = J^{-1}(\hat{\boldsymbol{d}}_i)\hat{D}(\hat{\boldsymbol{d}}_i),$$

column-vectors

$$\hat{\varphi}(\hat{d}_i) = \left\{ \hat{\varphi}_\beta(\hat{d}_i) \right\}_{\beta=1}^{6},$$

and coefficients $\gamma(\hat{d}_i) = \hat{\gamma}_i |J(\hat{d}_i)|$ such that

$$\bar{A}_{\alpha\beta} = \sum_{i=1}^{q} \hat{\gamma}_i \hat{z}(\hat{d}_i) = \sum_{i=1}^{q} \gamma(\hat{d}_i) \left(\hat{c}(\hat{d}_i) D_\beta(\hat{d}_i) \cdot D_\alpha(\hat{d}_i) \right.$$
$$\left. + \left(\hat{b}(\hat{d}_i) \cdot D_\beta(\hat{d}_i) + \hat{a}(\hat{d}_i)\hat{\varphi}_\beta(\hat{d}_i) \right) \hat{\varphi}_\alpha(\hat{d}_i) \right),$$

where $D_\alpha(\hat{d}_i)$ is the column of $D(\hat{d}_i)$ with index α. Since $x \cdot y = x^T y$, $xy^T = \{x_i y_j\}_{i,j=1}^{n}$, $x, y \in \mathbf{R}^n$, then local stiffness matrix can be expressed as

$$\bar{A} = \sum_{i=1}^{q} \gamma(\hat{d}_i) \left(\hat{c}(\hat{d}_i) D^T(\hat{d}_i) D(\hat{d}_i) \right.$$
$$\left. + \hat{\varphi}(\hat{d}_i) \left(\hat{b}(\hat{d}_i) D(\hat{d}_i) + \hat{a}(\hat{d}_i)\hat{\varphi}^T(\hat{d}_i) \right) \right).$$

The local forcing vector is calculated analogously, namely,

$$\bar{F} = \sum_{i=1}^{q} \gamma(\hat{d}_i) \hat{f}(\hat{d}_i) \hat{\varphi}(\hat{d}_i).$$

It is worth noting that the following is valid for the curvilinear triangular element (see 6.4)

$$\hat{f}(\hat{d}_i) = f(X\hat{\varphi}(\hat{d}_i)), \quad J(\hat{d}_i) = \hat{D}(\hat{d}_i)X^T, \quad X_{2\times 6} = (p_1, \dots, p_6),$$

where X is the matrix with interpolation nodes of the finite element τ.

Let us provide numerical values of coefficients $\gamma(\hat{d}_i)$ for two particular quadrature rules. First one contains three nodes ($q = 3$), it interpolates exactly polynomials of the second order ($\mathcal{G}_2$ mesh) and has the following nodes and coefficients:

$$\hat{d} = \begin{pmatrix} 0.5 & 0.5 & 0 \\ 0 & 0.5 & 0.5 \end{pmatrix}, \quad \hat{\gamma} = \frac{1}{6}(1 \quad 1 \quad 1).$$

Second one contains six points and interpolates exactly polynomials of the fourth order. Its nodes and coefficients are given by

$$a_1 = 0.09157621350977073, \quad a_2 = 0.09157621350977073,$$

$$b_1 = 0.4459484909159649, \quad b_2 = 0.4459484909159649,$$

$$a_3 = 1 - a_1 - a_2, \quad b_3 = 1 - b_1 - b_2,$$

$$\hat{d} = \begin{pmatrix} a_1 & a_1 & a_3 & b_1 & b_1 & b3 \\ a_2 & a_3 & a_2 & b_2 & b_3 & b2 \end{pmatrix},$$

$$\gamma_1 = 0.1099517436553218, \quad \gamma_2 = 0.2233815896780115,$$

$$\hat{\gamma} = \frac{1}{2} \begin{pmatrix} \gamma_1 & \gamma_1 & \gamma_1 & \gamma_2 & \gamma_2 & \gamma_2 \end{pmatrix}.$$

Now, we consider calculation of the boundary matrix $\bar{S} = \{\bar{S}_{\alpha\beta}\}$ and boundary vector $\bar{M} = \{\bar{M}_\alpha\}$ (see expression 5.8). Let e be the boundary edge with endpoints p_1, p_2 and midpoint p_3, and $\hat{e} = \{\hat{\theta} \in [0,1]\}$ is a basis edge having endpoints $\hat{p}_1 = 0$, $\hat{p}_2 = 1$ and midpoint $\hat{p}_3 = 0.5$. Then, basis functions defined on the edge can be expressed as

$$\hat{\varphi}_1(\hat{\theta}) = (1 - 2\hat{\theta})(1 - \hat{\theta}), \quad \hat{\varphi}_2(\hat{\theta}) = \hat{\theta}(2\hat{\theta} - 1), \quad \hat{\varphi}_3(\hat{\theta}) = 4\hat{\theta}(1 - \hat{\theta}).$$

Expression

$$x = x_e(\hat{\theta}) = p_1 \hat{\varphi}_1(\hat{\theta}) + p_2 \hat{\varphi}_2(\hat{\theta}) + p_3 \hat{\varphi}_3(\hat{\theta})$$

defines transformation $\hat{e} \to e$ ($\hat{p}_k \to p_k$, $k = 1, 2, 3$) and parametric representation of the edge as well. It is easy to see, that in case of straight line edge the midpoint $p_3 = (p_1 + p_2)/2$ and $x_e(\hat{\theta}) = p_1(1 - \hat{\theta}) + p_2\hat{\theta}$. Boundary integrals (5.8) for calculation of $\bar{S}_{\alpha\beta}$ and $\bar{M}_\alpha$ rewritten in terms of $\hat{\theta}$ over the basis edge $\hat{e}$ have the form of one-dimensional integrals

$$\bar{S}_{\alpha\beta} = \int_0^1 \sigma(x_e(\hat{\theta}))\hat{\varphi}_\alpha(\hat{\theta})\hat{\varphi}_\beta(\hat{\theta})l(\hat{\theta})d\hat{\theta},$$

$$\bar{M}_\alpha = \int_0^1 \mu(x_e(\hat{\theta}))\hat{\varphi}_\alpha(\hat{\theta})l(\hat{\theta})d\hat{\theta}, \tag{6.7}$$

where $l(\hat{\theta}) = \sqrt{(x_1'(\hat{\theta}))^2 + (x_2'(\hat{\theta}))^2}$. For the straight line edge, $l(\hat{\theta}) = |p_2 - p_1|$, and l is the length of edge.

To calculate one-dimensional integral, let us utilize quadrature rule

$$\int\limits_{0}^{1} \hat{z}(\hat{\theta})d\hat{\theta} \approx \sum_{i=1}^{q} \hat{\gamma}_i \hat{z}(\hat{d}_i).$$

Likewise calculation of two-dimensional integral (6.6), we introduce column-vectors $\hat{\boldsymbol{\varphi}}(\hat{d}_i) = \left\{\hat{\varphi}_\beta(\hat{d}_i)\right\}_{\beta=1}^{3}$, $d\hat{\boldsymbol{\varphi}}(\hat{d}_i) = \left\{\hat{\varphi}'_\beta(\hat{d}_i)\right\}_{\beta=1}^{3}$ and matrix of nodes coordinates $X_{2\times3} = (\boldsymbol{p}_1, \boldsymbol{p}_2, \boldsymbol{p}_3)$. Then, $x'(\hat{d}_i) = Xd\hat{\boldsymbol{\varphi}}(\hat{d}_i)$ and $l(\hat{d}_i) = |Xd\hat{\boldsymbol{\varphi}}(\hat{d}_i)|$ is the length of the vector $Xd\hat{\boldsymbol{\varphi}}(\hat{d}_i)$. Having introduced this, we come to the expressions for the matrix S and vector $\boldsymbol{M}$

$$S = \sum_{i=1}^{q} \gamma(\hat{d}_i)\sigma(X\hat{\boldsymbol{\varphi}}(\hat{d}_i))\hat{\boldsymbol{\varphi}}(\hat{d}_i)\hat{\boldsymbol{\varphi}}^T(\hat{d}_i),$$

$$\bar{\boldsymbol{M}} = \sum_{i=1}^{q} \gamma(\hat{d}_i)\mu(X\hat{\boldsymbol{\varphi}}(\hat{d}_i))\hat{\boldsymbol{\varphi}}(\hat{d}_i),$$

(6.8)

where $\gamma(\hat{d}_i) = \hat{\gamma}_i|Xd\hat{\boldsymbol{\varphi}}(\hat{d}_i)|$.

In conclusion, we provide coefficients for two particular quadrature rules defined on $[0, 1]$ which interpolates exactly polynomials of the third order. The first one is Gauss quadrature:

$$\hat{\gamma}_1 = \hat{\gamma}_2 = 0.5, \quad \hat{d}_1 = 0.5\left(1 - \sqrt{1/3}\right),$$

$$\hat{d}_2 = 0.5\left(1 + \sqrt{1/3}\right), \quad q = 2.$$

(6.9)

The second one is Simpson quadrature

$$\hat{\gamma}_1 = \hat{\gamma}_3 = 1/6, \quad \hat{\gamma}_2 = 4/6, \quad \hat{d}_1 = 0,$$

$$\hat{d}_2 = 0.5, \quad \hat{d}_3 = 1, \quad q = 3.$$

(6.10)

Using either (6.9) or (6.10) allows calculating boundary integrals (6.7) numerically and obtaining corresponding matrices (6.8).

Conclusion

I N this book we have considered creating computational framework for the linear finite element method (FEM) with application to solving ordinary and partial differential equations having mixed boundary conditions. Each step of the FEM algorithm is described in detail and programming codes in MATLAB and Python are composed. Although there exists various software packages implementing FEM, understanding the basics of the FEM algorithm is important to utilize these packages. Explanations of the FEM algorithm and computational framework given in this book not only help to understand basics of the method but to provide guidance of how to build customized code for solving a problem using FEM. The latter is important since it allows to develop coding skills and to become more flexible with solving any nonstandard problem where standard FEM packages could fail to solve (or simply these packages are not available). Apart from this, coding functions from this book can be extended or modified to build improved FEM framework with some specific features if required.

Of course, we have only considered simple case of linear FEM, but it could be starting point to build more sophisticated algorithms. For example, solving nonlinear PDE results in FEM system which requires applying numerical methods to solve nonlinear systems. Creating FEM programming codes for three-dimensional case could be the next step a reader starts to look at.

Variational Formulation of a Boundary Value Problem

H ERE we discuss variational approach to solve BVP. Variational methods are used in many applications where a problem can be expressed in terms of so-called variational principles. In accordance with one of these principles, function $u(x)$ being a solution to a BVP at the same time must be a stationary point of a variational functional

$$J(u) = \int_{s_0}^{s_1} F(x, u, u')dx,$$

where s_0 and s_1 are constants, $u(x)$ is a twice continuously differentiable function and function F is twice continuously differentiable with respect to its arguments. A stationary point is a function $u^*(x)$ which makes $J(u)$ stationary, or when the first-order change in $J(u)$ with respect to the function $u(x)$ vanishes. Usually, variational functional has a certain meaning, for example, it could be a potential energy and a stationary point is a function that minimizes the energy. It is known from variational calculus that a stationary point is solution to the Euler-Lagrange equation

$$-\frac{d}{dx}\frac{\partial F}{\partial u'} + \frac{\partial F}{\partial u} = 0. \tag{A.1}$$

If equation (A.1) is complemented with some boundary conditions

$$u(s_0) = u_{s_0}, \quad u(s_1) = u_{s_1}, \tag{A.2}$$

DOI: 10.1201/9781003265979-A

then we see that finding a stationary point of $J(u)$ is equivalent to solving BVP (A.1) – (A.2).

As an example, let us consider the functional

$$J(u) = \frac{1}{2} \int_{s_0}^{s_1} c(x) \left(u'(x)\right)^2 + a(x)u^2(x) - f(x)u(x)dx, \qquad \text{(A.3)}$$

where $c(x), a(x)$ and $f(x)$ are piecewise continuous functions and $c(x) > 0$, $b(x) > 0$. A variational problem is to find a function $u(x)$ which minimizes $J(u)$.

$$J(u) \to \min, \quad u(s_0) = u_{s_0}, \quad u(s_1) = u_{s_1}. \qquad \text{(A.4)}$$

Then, the Euler-Lagrange equation takes the form

$$- \left(c(x)u'(x)\right)' + a(x)u(x) = f(x). \qquad \text{(A.5)}$$

It can be shown that stationary point in this case is a function which minimizes (A.3). As a result, solving (A.3) – (A.4) is equivalent to solving BVP (A.5) – (A.2).

In practice, the BVP (A.5) – (A.2) is replaced by the integral expression (A.3) for which a minimization problem is considered with some approximation of unknown function. In the context of FEM, the unknown function u is represented through linear combination of some known functions, $\varphi_i(x)$, and coefficients, α_i, as follows:

$$u(x) = \sum_{i=1}^{N} \alpha_i \varphi_i(x). \qquad \text{(A.6)}$$

By substituting (A.6) into (A.3) we come to the problem of minimization of multivariate function $J(\alpha_i \varphi_i(x))$ with respect to variables α_i. According to the first-order necessary condition for a minimum, variables α_i can be found from equations

$$\frac{\partial}{\partial \alpha_i} J \left(\sum_{i=1}^{N} \alpha_i \varphi_i \right) = 0.$$

By adding to these equations conditions (A.2) we come to a system of equations for finding coefficients α_i which defines solution (A.6).

One limitation of the variational approach to solve BVP is that it is applicable only in case when a given differential equation represents Euler-Lagrange equation which is not always achievable. For example, BVP (1.1) is not allowing variational formulation. Instead, projective method can be used which is considered throughout this book.

Discussion of Global Interpolating Polynomial

L ET us consider a global approximating polynomial, $P_m(x)$, of the order m defined on the interval $[s_0, s_1]$ which interpolates a function $f(x)$. Set of interpolation nodes forms a sequence

$$s_0 = x_1 < x_2 < \cdots < x_{m+1} = s_1.$$

We want to analyse what happens when number of nodes increases (for example, to improve function approximation accuracy). Below, various types of approximation errors are considered, and it is shown how increasing number of interpolation nodes (and accordingly the degree of polynomial P_m) leads to accumulation of these errors.

Convergence with increasing number of nodes. Suppose that we have chosen some strategy of increasing polynomial degree, m, by introducing new nodes according to some rule. The question arises, whether it is always true that such polynomials converge to the source function, or

$$\Delta(P_m) = \max_{[s_0, s_1]} |f(x) - P_m(x)| \xrightarrow[m \to \infty]{} 0.$$

The short answer is not. To illustrate this, let us consider the Runge function[1] $f(x) = 1/(1 + 25x^2)$, $x \in [-1, 1]$, and a simple strategy which

---

[1]Given example is referred to Runge's phenomenon.

DOI: 10.1201/9781003265979-B

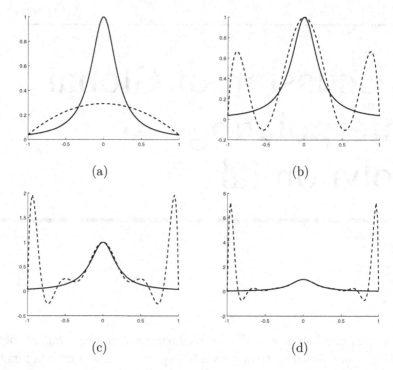

FIGURE B.1: Example of approximation divergence with increasing polynomial degree, m, for uniformly distributed interpolation nodes when (a) $m = 3$; (b) $m = 6$; (c) $m = 10$; (d) $m = 14$. Here, the source function is $f(x) = 1/(1 + 25x^2)$ (solid line), and approximation is performed using Lagrange basis functions (dashed line).

consists of having uniformly distributed nodes with the step $h = 2/m$, that is $x_i = -1 + (i - 1)h$, $i = 1, \ldots, m + 1$. The polynomial degree increases with increasing m and the result of applying this strategy with $m = \{6, 10, 14\}$ is shown in figure B.1. It can be seen that when m increases, approximation in the central part of the interval shows perfect result; however, polynomials diverge in the end regions, $0.73 \lesssim |x| \leq 1$.

We conclude that uniform distribution of nodes could potentially lead to approximation divergence. To understand why the node distribution is of importance, let us consider the approximation error, $\Delta(P_m)$, which can be estimated as

$$\Delta(P_m) \leq \frac{|f^{(m+1)}(\xi)|}{(m + 1)!} \max_{[s_0, s_1]} |\omega_{m+1}(x)|,$$

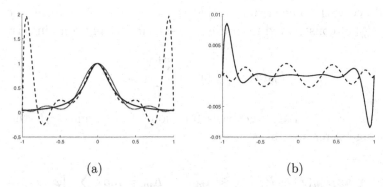

(a)   (b)

FIGURE B.2: Influence of the interpolation node distribution on the approximation error. Plot (a) shows the source function $f(x) = 1/(1 + 25x^2)$ (solid line), approximating polynomial of degree $m = 10$ having uniform node distribution (dashed line) and the polynomial which nodes are roots of the Chebyshev polynomial of the second kind (dotted line). Plot (b) illustrates dependence of the error function, ω_{m+1}, on the node distribution. Here, the solid line corresponds to the uniform distribution and the dashed line corresponds to the nodes being the Chebyshev polynomial roots.

where
$$\omega_{m+1}(x) = (x - x_1)(x - x_2) \cdots (x - x_{m+1}),$$

and $\xi \in [s_0, s_1]$. It can be seen that the error depends on function ω_{m+1} which, in turn, depends on the relative nodes location. The problem of divergence can be resolved by choosing nodes to be roots of an orthogonal polynomial, for example, taking nodes to be $x_i = -\cos(\pi(i-1)/m)$, $1 \leq i \leq m + 1$ improves convergence (see figure B.2).[2]

Although taking orthogonal polynomial roots as interpolation nodes is a good strategy, practically it could be inapplicable in some cases, for example, when interpolation nodes are prescribed through a table of function values which leaves no freedom of changing the nodes distribution.

Error of input data and approximation accuracy. Another source of polynomial approximation error arises from the fact that the function values at the interpolation nodes, $f_i = f(x_i)$, are specified or

---

[2]Nodes $x_i = -\cos(\pi(i-1)/m)$, $1 \leq i \leq m + 1$ are roots of the orthogonal polynomial of the form $(1-x^2)U'_{m-1}(x)$ where $U_{m-1}(x)$ is the Chebyshev polynomial of the second kind of order $m - 1$.

calculated with some errors, ϵ_i, that is $f_i^* = f_i + \epsilon_i$. As a result, polynomial $P_m^*(x)$ constructed from the nodes f_i^* inevitably contains errors as well

$$P_m(x) - P_m^*(x) = \sum_{j=1}^{m+1} \epsilon_j \varphi_j(x),$$

where $\varphi_j(x)$ is a Lagrange basis function. If $\epsilon^* = \max\{\epsilon_i\}$, then the following estimation takes place

$$\max_{[s_0,s_1]} |P_m(x) - P_m^*(x)| \leq \Lambda_m \epsilon^*, \quad \Lambda_m = \max_{[s_0,s_1]} \sum_{j=1}^{m+1} |\varphi_j(x)|.$$

Here, Λ_m is called the Lebesgue constant. It can be seen that the constant Λ_m represents the factor by which approximation error increases when the input data error is introduced. Sensitivity of the approximation method to the errors of input data is characterized by a conditional number which can be defined by the Lebesgue constant. The value of Λ_m depends only on nodes distribution but not on the interval length. Asymptotic estimation shows that the Lebesgue constant grows logarithmically with $m \to \infty$ if Chebyshev nodes are used. On the other hand, it grows exponentially in the case of equidistant nodes. The former fact leads to conclusion, that it is recommended to avoid using high order global interpolation polynomials having equidistant nodes distribution. Alternative approach is to use a piecewise local interpolation when the interval $[s_0, s_1]$ is split into subintervals and function f is approximated by interpolating polynomial of low order on each subinterval.

Interpolatory Quadrature Formulas

$\mathcal{Q}$UADRATURE formulas are used to approximate integrals of functions which are not computed exactly. The most common quadrature formulas look like

$$\int_{s_0}^{s_1} f\,dx \approx \sum_{i=1}^{n} \gamma_i f(x_i).$$

Here x_i are called nodes of the formula and γ_i are called weights.

Definition C.1 *Interpolatory quadrature is a type of quadrature rule when the integrand is approximated by another function, usually by interpolating polynomial, integral of which can be computed exactly.*

In other words, if $\varphi_i(x)$ are basis functions, then

$$\int_{s_0}^{s_1} f\,dx \approx \int_{s_0}^{s_1} \sum_{i=1}^{n} f(x_i)\varphi_i(x) = \sum_{i=1}^{n} f(x_i) \underbrace{\int_{s_0}^{s_1} \varphi_i(x)\,dx}_{\gamma_i} = \sum_{i=1}^{n} \gamma_i f(x_i).$$

Definition C.2 *The order of precision of a quadrature formula is the maximum degree of the polynomials which are integrated exactly by the formula.*

DOI: 10.1201/9781003265979-C

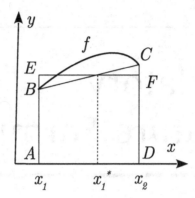

FIGURE C.1: Illustration of rectangle and trapezoid quadrature rules.

Obviously, if basis functions are integrated exactly then the order of precision of an interpolatory formula will be at least $n - 1$. Well-known rectangle formula, trapezoidal rule and Simpson rule are interpolatory quadrature rules corresponding to the polynomials of the zeroth, first and second orders, respectively.

It is seen from the above formula that the precision of quadrature formula can be increased by increasing number of interpolating nodes (degree of interpolating polynomial). However, if there is a freedom in choosing distribution of nodes then the precision could be increased through particular choice of nodes location with fixed number n. This can be illustrated as follows: Suppose we have a function f defined on the interval which is approximated by a linear polynomial with two nodes, x_1 and x_2 (see figure C.1). Then the trapezoid quadrature rule gives a value of the integral equalling the area $ABCD$ and this rule has the first order of precision. On the other hand, if we consider a rectangle formula with a freedom of choice of interpolation node, x_1^*, we can find location of x_1^* so that the area $AEFD$ (result of rectangle formula) would be equal to the area $ABCD$. In this situation, the rectangle formula results not in the zeroth order of precision but in the first order. This is achieved through the particular choice of node location.

Definition C.3 *An interpolatory quadrature rule having the highest order of precision with the given number of nodes is called Gaussian quadrature. Given n number of nodes, the highest precision order of $2n - 1$ can be reached.*

There are several algorithms for computing the nodes x_i and weights w_i of Gaussian quadrature rule and one of them is given below. We start by considering a quadrature for the interval $[-1, 1]$ with n nodes[1]

$$\int_{-1}^{1} f dx \approx \sum_{i=1}^{n} \gamma_i f(x_i).$$

Note that this formula is exact for the polynomial of some degree N if and only if it is exact for the set of functions $f = 1, x, x^2, \ldots, x^N$. This is equivalent to the fact that nodes and weights must satisfy the following system of nonlinear equations

$$\sum_{i=1}^{n} \gamma_i x_i^k = \int_{-1}^{1} x^k dx = \frac{1 - (-1)^{k+1}}{k+1}, \quad k = 0, \ldots, N.$$

It can be shown that the system has unique solution if the number of unknowns is equal to the number of equations, so $N = 2n - 1$. For example, Gaussian quadrature with two nodes corresponds to $n = 1$ and $N = 3$. Weights and nodes are found from the system

$$\gamma_1 + \gamma_2 = \int_{-1}^{1} dx = 2,$$

$$\gamma_1 x_1 + \gamma_2 x_2 = \int_{-1}^{1} x dx = 0,$$

$$\gamma_1 x_1^2 + \gamma_1 x_2^2 = \int_{-1}^{1} x^2 dx = \frac{2}{3},$$

$$\gamma_1 x_1^3 + \gamma_1 x_2^3 = \int_{-1}^{1} x^3 dx = 0.$$

---

[1]Integration over an arbitrary interval, $[s_0, s_1]$, can be obtained from the integration over the interval $[-1, 1]$ through the transformation $\tilde{x} = (s_0 + s_1)/2 + x(s_1 - s_0)/2$

$$\int_{s_0}^{s_1} f dx \approx \frac{s_1 - s_0}{2} \sum_{i=1}^{n} \gamma_i f(\tilde{x}_i).$$

By solving it, we find the values: $\gamma_1 = \gamma_2 = 1$, $x_1 = -1/\sqrt{3}, x_2 = 1/\sqrt{3}$. As a result we obtain Gaussian quadrature with two nodes being exact for polynomials of the third order

$$\int_{-1}^{1} f dx \approx f\left(-\frac{1}{\sqrt{3}}\right) + f\left(\frac{1}{\sqrt{3}}\right).$$

Nodes and weights for larger values of m can be found tabulated in many numerical analysis books.

Conditional number of interpolatory quadrature formulas. When computing integrals, an integrand is often known approximately with some error (for example, it could be computed with some rounding error or measured with error). As a result, instead of exact value of the integrand, f, approximated value is available, $f^* = f + \epsilon$, having some error ϵ. Let us estimate sensitivity of interpolatory quadrature with respect to input data, or equivalently, defined a quadrature conditional number, ν. If we assume that $|f(x) - f^*(x)| < \epsilon^* \; \forall x \in [s_0, s_1]$, then

$$\left| \sum_{i=1}^{n} \gamma_i f(x_i) - \sum_{i=1}^{n} \gamma_i f^*(x_i) \right| \le \epsilon^* \sum_{i=1}^{n} |\gamma_i| = \nu \epsilon^*, \quad \nu = \sum_{i=1}^{n} |\gamma_i|.$$

As such, perturbing input data lead to increasing quadrature error by factor ν which is a conditional number. One important observation regarding the conditional number is that

$$\nu = \sum_{i=1}^{n} |\gamma_i| \ge \sum_{i=1}^{n} \gamma_i,$$

which means that if all weights are positive we obtain the smallest conditional number; otherwise, we come to an ill-conditional quadrature. All weights of Gaussian quadrature are positive which guarantees stability with respect to input data errors.

Quadrature Rules and Orthogonal Polynomials

IN general case, Gaussian quadrature rule for integral numerical calcu- lation can be derived using property of orthogonal polynomials. Here, we provide some basics of the orthogonal polynomials theory and give generic formulas to calculate quadrature nodes and weights. Let us start with a definition.

Definition D.1 *A family of polynomials* $\{P_n(x)\}_{n=0}^{\infty}$, $x \in (s_0, s_1)$, *is called a sequence of orthogonal polynomials with a weight* $\omega(x)$ *if the following holds*

$$P_m(x) \in \mathcal{G}_m, \quad \int_{s_0}^{s_1} \omega(x)P_i(x)P_j(x)dx = 0, \quad i \neq j.$$

It is easy to see that orthogonal polynomials are defined up to con- stant multipliers. Choice of multiplier values can be dictated by some normalization, for example, they can be chosen so that to obtain unity leading polynomial coefficient. An overview of the most important prop- erties of the orthogonal polynomials is given below.

1. The polynomial $P_n(t)$ has n real distinctive roots which are within the interval of orthogonality.

2. The sequence of orthogonal polynomials satisfies a recurrence re- lation of the form

$$P_{k+1}(x) = (a_k x - b_k)P_k(x) - c_k P_{k-1}(x), \quad k \geq 0, \quad P_{-1}(x) = 0.$$

DOI: 10.1201/9781003265979-D

3. Polynomial roots, $\{x_k\}_{k=1}^m$, are the eigenvalues of the following tridiagonal matrix

$$
A_n = \begin{pmatrix}
\alpha_0 & \sqrt{\beta_1} & & & \\
\sqrt{\beta_1} & \alpha_1 & \sqrt{\beta_2} & & \\
& \ddots & \ddots & \ddots & \\
& & \sqrt{\beta_{n-2}} & \alpha_{n-2} & \sqrt{\beta_{n-1}} \\
& & & \sqrt{\beta_{n-1}} & \alpha_{n-1}
\end{pmatrix},
\tag{D.1}
$$

where

$$
\alpha_k = \frac{b_k}{a_k}, \quad k \geq 0, \quad \beta_k = \frac{c_k}{a_{k-1}a_k}, \quad k \geq 1.
\tag{D.2}
$$

4. Polynomial roots, $\{x_k\}_{k=1}^n$, are the nodes of Gauss Jacobi quadrature

$$
\int_{s_0}^{s_1} f(x)\omega(x)dx \approx \sum_{k=1}^n \gamma_k f(x_k),
\tag{D.3}
$$

where all $\gamma_k > 0$ and the quadrature is exact if f is a polynomial of order $2n - 1$. Vector of quadrature weights, $\gamma = \{\gamma_k\}_{k=1}^m$, is found from the expression

$$
\gamma_k = U_{1k}^2 m_0, \quad m_0 = \int_{s_0}^{s_1} \omega(x)dx,
\tag{D.4}
$$

where U_{1k} is the first row of the matrix U, columns of which are normalized eigenvectors (i.e., eigenvectors with unity Euclidean norm) of the matrix A_n, and $A_n = U\Lambda U^{-1}$, $\Lambda = \text{diag}(x_1, \ldots, x_n)$.

An important family of orthogonal polynomials are Jacobi polynomials determined by the weight function of the form

$$
\omega(x) = (1 - x)^\alpha (1 + x)^\beta, \quad (s_0, s_1) = (-1, 1), \quad \alpha, \beta > -1.
$$

In this case, orthogonal polynomials depend on parameters $\{\alpha, \beta\}$ and first two polynomials can be expressed as

$$
P_0^{(\alpha,\beta)}(x) = 1, \quad P_1^{(\alpha,\beta)}(x) = \frac{1}{2}\left((\alpha + \beta + 2)x + \alpha - \beta\right).
$$

Coefficients of recurrent formula are found through

$$a_k = \frac{(2k + \alpha + \beta + 1)(2k + \alpha + \beta + 2)}{2(k+1)(k + \alpha + \beta + 1)}, \tag{D.5}$$

$$b_k = \frac{(\beta^2 - \alpha^2)(2k + \alpha + \beta + 1)}{2(k+1)(k + \alpha + \beta + 1)(2k + \alpha + \beta)},$$

$$c_k = \frac{(k + \alpha)(k + \beta)(2k + \alpha + \beta + 2)}{(k+1)(k + \alpha + \beta + 1)(2k + \alpha + \beta)}.$$

Integral in (D.4) can be found as follows:

$$m_0 = 2^{\alpha+\beta+1} \frac{\Gamma(\alpha + 1)\Gamma(\beta + 1)}{\Gamma(\alpha + \beta + 2)}, \tag{D.6}$$

where $\Gamma(x)$ is a gamma function which is defined as $\Gamma(x) = (x - 1)!$ for any positive integer x. Derivatives of polynomials can be determined through the expression

$$\frac{d}{dx} P_n^{(\alpha,\beta)} = \frac{1}{2}(m + \alpha + \beta + 1) P_{n-1}^{(\alpha+1,\beta+1)}(x). \tag{D.7}$$

Let us indicate some particular cases. When $\alpha = \beta = -1/2$ we obtain Chebyshev polynomials of the first type, T_n. Parameters $\alpha = \beta = 0$ correspond to Legendre polynomials, $L_n = P_n^{(0,0)}$, which represent an important class of polynomials for obtaining quadrature rules. It is worth indicating that when considering Legendre polynomials, we have $\omega(x) = 1$ and the quadrature (D.3) transforms into Gaussian quadrature rule over the interval $[-1, 1]$

$$\int_{-1}^{1} f(x)dx \approx \sum_{k=1}^{n} \gamma_k f(x_k),$$

By substituting $\alpha = 0$ and $\beta = 0$ into (D.5) and (D.6) we find the matrix coefficients (D.2) to be $\alpha_k = 0$, $\beta_k = k^2/(4k^2 - 1)$, and the integral $m_0 = 2$. As a result, quadrature weights to be $\gamma_k = 2U_{1k}^2$ and corresponding quadrature nodes are the eigenvalues of matrix A_n.

The nodes of the Gaussian quadrature, being the zeros of Legendre polynomials, do not encompass the boundaries -1 and 1 of the interval $[-1, 1]$. In some circumstance, it is desirable to include these points in the boundaries. This is possible at the price of reducing by 2 units the degree of exactness of the Gaussian quadrature which results in Lobatto quadrature

$$\int_{-1}^{1} f(x)dx \approx \frac{2}{n(n-1)}f(-1) + \sum_{k=2}^{n-1} \gamma_k f(x_k) + \frac{2}{n(n-1)}f(1).$$

Similar to the Gaussian quadrature problem stated previously, in Lobatto quadrature we constrain $x_1 = -1$ and $x_n = 1$, which means we now have less degree of freedom and the Lobatto quadrature is exact for polynomials of order $2m - 3$. In this case, quadrature nodes are roots of the polynomial $(1 - x^2)L'_{n-1}(x)$. Indeed, x_1 and x_n are the roots of $(1 - x^2)$, and nodes $x_2, \ldots, x_{n-1}$, are the $n - 2$ zeros of the polynomial L'_{n-1}, or equivalently the points where the polynomial L_{n-1} is extremal. From (D.7) we see that roots of L'_{n-1} are roots of $P_{n-2}^{(1,1)}$, and by substituting $\alpha = 1$ and $\beta = 1$ into (D.5) and (D.6) we find the matrix coefficients (D.2) to be $\alpha_k = 0$, $\beta_k = k(k + 1)/((2k + 1)(2k + 3))$, and the integral $m_0 = 4/3$. Once quadrature weights, $\tilde{\gamma}_k$, for L'_{n-1} are found, corresponding weights for $(1 - x^2)L'_{n-1}$ are defined as $\gamma_k = \tilde{\gamma}_k/(1 - x_k^2)$.

Computational Framework in Python

E.1 ASSEMBLING MATRICES

How to calculate and assemble FEM matrices is explained in sections 5.3–5.5 where MATLAB function `assemba()` for the vectorized version of the assembling algorithm is given. It is straightforward to translate this function into Python with some minor changes.

Below we provide Python code for matrix calculating and assembling algorithm utilizing sparse matrices. It follows the structure of the MATLAB code given in section 5.5.

```python
1  import numpy
2  import scipy.sparse as sparse
3  def assemblingAF(p,t,c,a,b1,b2,f=None):
4      # assemblingAF - vectorized algorithm of the assembling
5      #stiffness matrix A and forcing vector F for the G1 mesh.
6      # Stiffness matrix A is associated with the PDE operator
7      # -div(c*grad(u)) + b*grad(u)+a*u
8      #The following call is allowed :
9      # A=assemblingAF(p,t,c,a,b1,b2)
10     #
11     np=p.shape[1]
12     # mesh point indices
13     k1=t[0,:]
14     k2=t[1,:]
15     k3=t[2,:]
```

DOI: 10.1201/9781003265979-E

```
16    sdl=t[3,:] # subdomain labels
17    # barycenter of the triangles
18    x1=(p[0,k1]+p[0,k2]+p[0,k3])/3
19    x2=(p[1,k1]+p[1,k2]+p[1,k3])/3
20    # gradient of the basis functions, multiplied by J
21    g1_x1=p[1,k2]—p[1,k3]
22    g1_x2=p[0,k3]—p[0,k2]
23    g2_x1=p[1,k3]—p[1,k1]
24    g2_x2=p[0,k1]—p[0,k3]
25    g3_x1=p[1,k1]—p[1,k2]
26    g3_x2=p[0,k2]—p[0,k1]
27    J=abs(g3_x2*g2_x1—g3_x1*g2_x2)  # J=2*area
28    # evaluate c , b , a on triangles barycenter
29    cf=c(x1,x2,sdl)
30    af=a(x1,x2,sdl)
31    b1f=b1(x1,x2,sdl)
32    b2f=b2(x1,x2,sdl)
33    # diagonal and off diagonal elements of mass matrix
34    ao=(af/24)*J
35    ad=4*ao  # 'exact' integration
36    # ao=(af/18).*J ; ad=3*ao ; # quadrature rule
37    # coefficients of the stiffness matrix
38    cf =(0.5*cf)/J
39    a12=cf*(g1_x1*g2_x1+g1_x2*g2_x2)+ao
40    a23=cf*(g2_x1*g3_x1+g2_x2*g3_x2)+ao
41    a31=cf*(g3_x1*g1_x1+g3_x2*g1_x2)+ao
42    A=[]
43    F=[]
44    if all(b1f==0) and all(b2f==0): # symmetric problem
45        A=sparse.coo_matrix((a12,(k1,k2)),shape=(np,np))
46        A=A+sparse.coo_matrix((a23,(k2,k3)),shape=(np,np))
47        A=A+sparse.coo_matrix((a31,(k3,k1)),shape=(np,np))
48        A=A+sparse.coo_matrix.transpose(A)
49        A=A+sparse.coo_matrix((ad—a31—a12,(k1,k1)),shape=(np,np))
50        A=A+sparse.coo_matrix((ad—a12—a23,(k2,k2)),shape=(np,np))
51        A=A+sparse.coo_matrix((ad—a23—a31,(k3,k3)),shape=(np,np))
52    else:
53        # b contributions
54        b1f=b1f/6
55        b2f=b2f/6
```

```
56        bg1=b1f*g1_x1+b2f*g1_x2
57        bg2=b1f*g2_x1+b2f*g2_x2
58        bg3=b1f*g3_x1+b2f*g3_x2
59        A=sparse.coo_matrix((a12+bg2,(k1,k2)),shape=(np,np))
60        A=A+sparse.coo_matrix((a23+bg3,(k2,k3)),shape=(np,np))
61        A=A+sparse.coo_matrix((a31+bg1,(k3,k1)),shape=(np,np))
62        A=A+sparse.coo_matrix((a12+bg1,(k2,k1)),shape=(np,np))
63        A=A+sparse.coo_matrix((a23+bg2,(k3,k2)),shape=(np,np))
64        A=A+sparse.coo_matrix((a31+bg3,(k1,k3)),shape=(np,np))
65        A=A+sparse.coo_matrix((ad-a31-a12+bg1,(k1,k1)), \
66                           shape=(np,np))
67        A=A+sparse.coo_matrix((ad-a12-a23+bg2,(k2,k2)), \
68                           shape=(np,np))
69        A=A+sparse.coo_matrix((ad-a23-a31+bg3,(k3,k3)), \
70                           shape=(np,np))
71     if f!=None:
72        ff=f(x1,x2,sdl)
73        ff=(ff/6)*J
74        k_zero=numpy.zeros(k1.shape)
75        F=sparse.coo_matrix((ff,(k1,k_zero)),shape=(np,1))
76        F=F+sparse.coo_matrix((ff,(k2,k_zero)),shape=(np,1))
77        F=F+sparse.coo_matrix((ff,(k3,k_zero)),shape=(np,1))
78        return A,F
79     else:return A
```

Listing E.1: Python code for assembling stiffness matrices.

E.2 ASSEMBLING BOUNDARY CONDITIONS

Implementation of boundary condition calculation in Python is based on the approach described in section 5.6. We use dictionary structure containing list of boundary segments and corresponding function names defined on these segments. For example, by defining the function uD(x1,x2,sdl)

```
1 def uD(x1,x2,sdl):
2     f=numpy.zeros(x1.shape)
3     # segments in local numeration:
4     # 1-st segment
5     I=sdl==0
6     f[I]=numpy.square(x1[I])+numpy.square(x2[I])
```

```
7    # 2-nd segment
8    I=sdl==1
9    f[I]=numpy.square(x1[I])+numpy.square(x2[I])
10   return f
```

Listing E.2: Python code for defining BC.

we can specify Dirichlet boundary conditions on the first and third segments as

```
bc={}
bc["bsD"]=[1, 3]
bc["uD"]=uD
```

Python implementation of the function `assemblingBC()` given below performs assembling boundary conditions and repeat the logic of the corresponding MATLAB function.

```
1  import numpy
2  import scipy.sparse as sparse
3  def assemblingBC(bc,p,e,nargout):
4      # assemblingBC - assembling boundary conditions
5      # The following call is also allowed :
6      # N,S = assemblingBC(bc,p,e,nargout)
7      # N,S,UD = assemblingBC(bc,p,e,nargout)
8      np=p.shape[1]
9      S=[]
10     M=[]
11     N=sparse.eye(np)
12     UD=[]
13     if (bc["bsR"]==None) and \
14        (bc["sg"]==None) and \
15        (bc["mu"]==None):
16         return # all boundaries are homogeneous Robin b.c.
17     e=e[:,numpy.logical_or(e[5,:]==0,e[6,:]==0)] #boundary edges
18     k=e[4,:]
19     bsg=numpy.transpose(numpy.nonzero( \
20                      sparse.coo_matrix( \
21            (numpy.ones(len(k)),(k,k)),shape=(np,np))))[:,0]
22     nbsg=len(bsg)
23     # set local numeration of boundary segments
24     local=numpy.zeros((nbsg))
25     # set mixed boundary conditions
```

```
26      if nargout>=2:
27          bsR=bc["bsR"]
28          if len(bsR)>0:
29              if bsR==None: # all BC are mixed
30                  bsR=bsg
31              local[bsR]=range(0,len(bsR))
32              # find boudary edges with mixed b.c.
33              eR=e[:,numpy.isin(k,bsR)]
34              eR=eR[[0, 1, 4],:]
35              k1=eR[0,:]
36              k2=eR[1,:] # indices of starting points and endpoints
37              x1=0.5*(p[0,k1]+p[0,k2])
38              x2=0.5*(p[1,k1]+p[1,k2])
39              h=numpy.sqrt(numpy.square(p[0,k2]-p[0,k1])
40                  +numpy.square(p[1,k2]-p[1,k1])) # edges length
41              sdl=local[eR[2,:]]
42              # evaluate sigma , mu on edges barycenter
43              sf=bc["sg"](x1,x2,sdl)
44              mf=bc["mu"](x1,x2,sdl)
45              # 'exact' integration
46              so=(sf/6)*h
47              sd=2*so
48              # quadrature rule
49              #so=(sf/4)*h
50              #sd = so
51              S=sparse.coo_matrix((so,(k1,k2)),shape=(np,np))
52              S=S+sparse.coo_matrix((so,(k2,k1)),shape=(np,np))
53              S=S+sparse.coo_matrix((sd,(k1,k1)),shape=(np,np))
54              S=S+sparse.coo_matrix((sd,(k2,k2)),shape=(np,np))
55              if nargout==4:
56                  mf=bc["mu"](x1,x2,sdl)
57                  mf=(mf/2)*h
58                  k_zero=numpy.zeros(k1.shape)
59                  M=sparse.coo_matrix((mf,(k1,k_zero)), \
60                                      shape=(np,1))
61                  M=M+sparse.coo_matrix((mf,(k2,k_zero)), \
62                                      shape=(np,1))
63      # set Dirichlet boundary conditions
64      bsD=bc["bsD"]
65      if len(bsD)>0:
```

```
66    if bsD==None: # all b.c. are Dirichlet
67        bsD=bsg
68    local[bsD]=range(0,len(bsD))
69    if all(local==0):
70        print('error. bsD+bsR~=number of boundary segments')
71    eD=e[:,numpy.isin(k,bsD)] # boudary with Dirichlet BC
72    eD=eD[[0, 1, 4],:]
73    sdl=local[eD[2,:]]
74    # indices of start and end points
75    k1=eD[0,:]
76    k2=eD[1,:]
77    iD=numpy.concatenate([k1, k2]) # Dirichlet points indices
78    i_d=numpy.transpose(numpy.nonzero(
79        sparse.coo_matrix(
80            (numpy.ones(len(iD)),(iD,iD)), \
81            shape=(np,np))))[:,0]
82    # iN - indices of nonDirichlet points
83    iN=numpy.ones((np))
84    iN[i_d]=numpy.zeros((len(i_d)))
85    iN=numpy.transpose(numpy.nonzero(iN)).flatten()
86    niN=len(iN)
87    N=sparse.coo_matrix((numpy.ones(niN), \
88                (iN,range(0,niN))),shape=(np,niN))
89    if nargout>=3: # evaluate UD on Dirichlet points
90        UD=sparse.csr_matrix((1,np))
91        UD[0,k1]=bc["uD"](p[0,k1],p[1,k1],sdl)
92        UD[0,k2]=bc["uD"](p[0,k2],p[1,k2],sdl)
93        UD=UD.reshape((np,1))
94 if nargout==2:return N,S
95 elif nargout==3:return N,S,UD
96 elif nargout==4:return N,S,UD,M
97 else:print('assemblingBC: Wrong number of output parameters.')
```

Listing E.3: Python code for assembling BC.

Here, we need to explain how to implement in Python some MAT-LAB functions. For instance, MATLAB function find(a) which finds indices and values of nonzero elements of the array a can be represented in Python as function numpy.nonzero(a) which returns a tuple of arrays, one for each dimension of a, containing the indices of the nonzero elements in that dimension. To group the indices by element, rather than dimension, we can use `numpy.transpose(numpy.nonzero(a))` which results in two-dimensional array of indices with a row for each nonzero element. Logical operations over Boolean arrays can be performed through special functions of the `numpy` package, for example, function `numpy.logical_or(x1,x2)` applies logical OR to the elements of `x1` and `x2`. MATLAB function `ismember(x1,x2)` corresponds to Python function `numpy.isin(x1,x2)` which returns a Boolean array of the same shape as array `x1` that is `True` where an element of `x1` is in `x2` and `False` otherwise. Listing E.4 shows a function `assemblingPDE()` code for solving a PDE problem. Function `spsolve()` is used here to solve linear system of sparse matrices.

```
1  import numpy
2  import scipy.sparse as sparse
3  from scipy.sparse.linalg import spsolve
4  def assemblingPDE(bc,p,e,t,c,a,b1,b2,f,nargout):
5      # A,F,N,UD,S,M=assemblingPDE(bc,p,e,t,c,a,b1,b2,f,nargout) -
6      #returns FEM matrices of the PDE problem.
7      # u=assemblingPDE(bc,p,e,t,c,a,b1,b2,f,nargout) -
8      #returns solution to the PDE problem
9      A=[]; F=[] ; UD=[]; S=[]; M=[]
10     if nargout==6:
11         A,F=assemblingAF(p,t,c,a,b1,b2,f)
12         N,S,UD,M=assemblingBC(bc,p,e,4)
13         return A,F,N,UD,S,M
14     elif (nargout == 1):
15         A,F=assemblingAF(p,t,c,a,b1,b2,f)
16         N,S,UD,M=assemblingBC(bc,p,e,4)
17         if N.shape[1]==p.shape[1]: # no Dirichlet BC
18             A=A+S; F=F+M
19         else:
20             Nt=sparse.coo_matrix.transpose(N)
21             A=A+S
22             F=Nt @ ((F+M)—A @ UD)
23             A=Nt @ A @ N
24         u=sparse.linalg.spsolve(A,F)
25         u=(sparse.coo_matrix(u)).reshape((u.shape[0],1))
26         return (N @ u+UD)
27     else: print('assembpde: Wrong number of output parameters.')
```

Listing E.4: Python code for solving PDE.

E.3 SOLVING EXAMPLE PROBLEM

As a final step, let us create a Python code for solving the example problem formulated in section 5.8. The code is self-documented and easy to read.

```python
1  import pygmsh
2  import numpy
3  import scipy.sparse as sparse
4  from scipy.sparse.linalg import spsolve
5  #geometry definition
6  def disk(h):
7      geom = pygmsh.built_in.Geometry()
8      # cirlcle
9      cirlcle_p0 = geom.add_point([0.0, 0.0, 0.0], h)#center
10     cirlcle_p1 = geom.add_point([-1.0, 0.0, 0.0], h)
11     cirlcle_p2 = geom.add_point([0.0, -1.0, 0.0], h)
12     cirlcle_p3 = geom.add_point([1.0, 0.0, 0.0], h)
13     cirlcle_p4 = geom.add_point([0.0, 1.0, 0.0], h)
14
15     arc_1=geom.add_circle_arc(cirlcle_p1, cirlcle_p0, cirlcle_p2)
16     arc_2=geom.add_circle_arc(cirlcle_p2, cirlcle_p0, cirlcle_p3)
17     arc_3=geom.add_circle_arc(cirlcle_p3, cirlcle_p0, cirlcle_p4)
18     arc_4=geom.add_circle_arc(cirlcle_p4, cirlcle_p0, cirlcle_p1)
19
20     circle = geom.add_line_loop([arc_1,arc_2,arc_3,arc_4])
21     disk = geom.add_plane_surface(circle)
22     #make boundary and area physical
23     geom_labels={}
24     group_id=geom._TAKEN_PHYSICALGROUP_IDS
25     ph_disk = geom.add_physical_surface(disk,label="1")
26     geom_labels[group_id[-1]]="1"
27     ph_arc_1 = geom.add_physical_line(arc_1,label="1:L1:R0")
28     geom_labels[group_id[-1]]="1:L1:R0"
29     ph_arc_2 = geom.add_physical_line(arc_2,label="2:L1:R0")
30     geom_labels[group_id[-1]]="2:L1:R0"
31     ph_arc_3 = geom.add_physical_line(arc_3,label="3:L1:R0")
32     geom_labels[group_id[-1]]="3:L1:R0"
33     ph_arc_4 = geom.add_physical_line(arc_4,label="4:L1:R0")
34     geom_labels[group_id[-1]]="4:L1:R0"
35     return geom,geom_labels
```

```
36
37 # boundary conditions
38 def uD(x1,x2,sdl):
39     f=numpy.zeros(x1.shape)
40     I=sdl==0
41     f[I]=numpy.square(x1[I])+numpy.square(x2[I])
42     I=sdl==1
43     f[I]=numpy.square(x1[I])+numpy.square(x2[I])
44     return f
45
46 def sigma(x1,x2,sdl):
47     f=numpy.zeros(x1.shape)
48     I=sdl==0
49     f[I]=0
50     I=sdl==1
51     f[I]=2
52     return f
53
54 def mu(x1,x2,sdl):
55     f=numpy.zeros(x1.shape)
56     I=sdl==0
57     f[I]=2*(2+x1[I]+x2[I])*(numpy.square(x1[I])+ \
58                            numpy.square(x2[I]))
59     I=sdl==1
60     f[I]=2*(2+x1[I]+x2[I]+1)*(numpy.square(x1[I])+ \
61                              numpy.square(x2[I]))
62     return f
63
64 #main function
65 def testPDE():
66     np=[] ; nt=[] ; errh2=[] ; err=[] ;
67     for h in [0.5, 0.1, 0.05, 0.02]:
68         geom,geom_labels=mg.disk(h)
69         p,t,e = generateMesh(geom,geom_labels)
70         p=numpy.transpose(p)
71         t=numpy.transpose(t)
72         e=numpy.transpose(e)
73         x1=p[0,:]
74         x2=p[1,:]
75         def exact(x1,x2,sdl):
```

```
76              return numpy.square(x1)+numpy.square(x2)
77          def c(x1,x2,sdl): return 2+x1+x2
78          def a(x1,x2,sdl): return x1+x2
79          def f(x1,x2,sdl):
80              return −8−6*(x1+x2)+(2+x1+x2)* \
81                      (numpy.square(x1)+numpy.square(x2))
82          def b1(x1,x2,sdl): return x1
83          def b2(x1,x2,sdl): return x2
84          u_exact=numpy.transpose(exact(x1,x2,1))
85          bc={}
86          bc["bsD"]=[1, 3]
87          bc["uD"]=uD
88          bc["bsR"]=[4, 2]
89          bc["sg"]=sigma
90          bc["mu"]=mu
91          A,F=assemblingAF(p,t,c,a,b1,b2,f)
92          N,S,UD,M=assemblingBC(bc,p,e,4)
93          u=assemblingPDE(bc,p,e,t,c,a,b1,b2,f,1).toarray()
94          u_exact=u_exact.reshape(u.shape)
95          np.append(p.shape[1])
96          nt.append(t.shape[1])
97          h2=h*h
98          err.append(numpy.linalg.norm(u_exact−u,numpy.inf))
99          errh2.append(numpy.linalg.norm(u_exact−u,numpy.inf)/h2)
100     print('{:<10s}{:<10s}{:^10s}{:^10s}'.format(
101             "np","nt","err","errh2"))
102     print()
103     for i in range(len(err)):
104         print('{:<10d}{:<10d}{:^10.5f}{:^10.5f}'.format(
105                 np[i],nt[i],err[i],errh2[i]))
```

Listing E.5: Python code for solving example problem.

Here, geometry is defined in function disk() and BC are specified in functions uD(), sigma() and mu(). Function testPDE() when called calculates solution values and prints solution error in various norms.

Bibliography

[1] G. Dhatt, G. Touzot, and E. Lefrancois. *Finite Element Method*. London, UK: ISTE Ltd, 2012.

[2] A.J.M. Ferreira. *MATLAB Codes for Finite Element Analysis*. Vol. 157. Solid Mechanics and Its Applications. Springer, 2009.

[3] Mark S. Gockenbach. *Understanding and Implementing the Finite Element Method*. Society for Industrial and Applied Mathematics, 2006.

[4] E. Jones, T. Oliphant, P. Peterson, et al. *SciPy: Open Source Scientific Tools for Python*. 2001. URL: http://www.scipy.org/.

[5] Gang Li. *Introduction to the Finite Element Method and Implementation with MATLAB*. Cambridge University Press, 2020.

[6] *MATLAB and Partial Differential Equation Toolbox Release 2019a*. Natick, Massachusetts, USA, 2019. URL: http://www.mathworks.com/.

[7] Travis E. Oliphant. *A Guide to NumPy*. USA: Trelgol Publishing, 2006.

Index

Printed in the United States
by Baker & Taylor Publisher Services

Printed in the United States
by Baker & Taylor Publisher Services